D1062369

TJ   80286
825   Eldridge, Frank R.
.E4   Wind machines
1980

DATE			
APR 5 1990	AP 12 11		
SEP 2 6 1991	NV 22 12		
NOV 1 8 1991			
APR 23 '92			
DEC 0 4 '92			
MAY 5 '94			
AUG 1 0 '95			
fac			
DE 17 '09			
MR 25 09			
AP 14 '10			

WIND MACHINES

second edition

WIND MACHINES

second edition

Frank R. Eldridge

Metrek Division
The MITRE Corporation

The MITRE Energy Resources and Environment Series

 VAN NOSTRAND REINHOLD COMPANY

NEW YORK   CINCINNATI   ATLANTA   DALLAS   SAN FRANCISCO
LONDON   TORONTO   MELBOURNE

Van Nostrand Reinhold Company Regional Offices:
New York Cincinnati Atlanta Dallas San Francisco

Van Nostrand Reinhold Company International Offices:
London Toronto Melbourne

Library of Congress Catalog Card Number: 79-22353
ISBN: 0-442-26134-9

Manufactured in the United States of America

Published by Van Nostrand Reinhold Company
135 West 50th Street, New York, N.Y. 10020

Published simultaneously in Canada by Van Nostrand Reinhold Ltd.

15 14 13 12 11 10 9 8 7 6 5 4 3 2 1

Library of Congress Cataloging in Publication Data

Eldridge, Frank R.
   Wind machines.

   Bibliography: p.
   Includes index.
   1. Windmills. 2. Wind power. I. Title.
TJ825.E4   1979   621.4'5   79-22353
ISBN 0-442-26134-9

To my wife
Peg

PREFACE

The purpose of this book is to provide a survey of the present status of the viability, history, taxonomy, and future potential of various types and sizes of wind machines that might be used to help meet future United States energy demands. Also discussed are various possible applications of wind machines, as well as siting problems, performance characteristics, and system designs for such machines.

It should be emphasized that wind is a clean, replenishable source of energy, and though it is intermittent and relatively dilute in nature as compared to fossil and nuclear fuels, it constitutes a large, practically untapped energy resource.

There are many possible ways of extracting useful energy from the wind. For instance, the mechanical power derived from wind turbines, which might vary in size from several feet to several hundreds of feet in diameter, can be used to drive electrical generators, or pump water or air, or perform other useful work. The largest anticipated wind-powered units of current design are small by utility company standards. They are limited to an electrical generating capacity of a few megawatts and, if inter-tied with utility grids, will require extensive electrical transmission facilities and proper phasing of the current generated. The economic viabilities of all these systems are strongly dependent not only on such factors as system reliabilities, lifetimes, costs, and prices of energy derived from competing energy sources, but also on the magnitude and time distribution of wind velocities at the specific geographical locations chosen for particular applications. The magnitude of the wind velocity at an installation site is particularly important, since the wind system power output varies in proportion to the cube of the wind velocity.

Current research on new system concepts that use wind concentrators, diffusers, and vortex generators indicate that such systems may offer the possibility of increasing both the ambient wind speeds and the resulting power output of a given size wind turbine by very significant amounts. In some cases, such designs might be included in the future, as integral parts of multi-purpose buildings to improve the aesthetics of wind systems and to lower the cost of the generation of power from the wind.

A glossary of commonly used words and phrases is included in the Appendix. All costs and prices are given in terms of 1978 dollars, unless otherwise stated.

FRANK R. ELDRIDGE

ACKNOWLEDGMENTS

This book is a revision of a monograph that was originally sponsored by the National Science Foundation and first published in October 1975. Since then, the monograph has gone through two further printings without change.

In the present edition, materials have been added which describe new developments in the design and application of wind machines and associated storage systems that have taken place since the first edition was published. In particular, more emphasis has been given to small wind machines used for dispersed applications.

The cover was created by Robert Berks, the distinguished sculptor and artist who is noted for his head of the late President Kennedy, located in the Kennedy Center in Washington, D.C. and his statue of Albert Einstein on the grounds of the National Academy of Sciences in the same city. The graphic arts work has been produced primarily by William Baumgardner and Jane Andrle of MITRE.

The information presented in this survey has been extracted from a large number of sources, as indicated in the appended bibliography and the list of picture credits. I am very grateful to the many authors and photographers from whom these selected materials were drawn. In particular, I would like to express my appreciation to Louis Divone of the Wind Energy Branch of the U.S. Department of Energy, who commissioned the 1975 monograph and who provided basic guidance in outlining and structuring that document; to George Tennyson and Donald Teague of DOE; to Marcel Harper of the National Science Foundation, who was Program Manager for the effort that produced the original monograph; to Olle Ljungstrom of the National Swedish Board for Energy Source Development; to Richard Greeley, Willis Jacobsen, and Grant Miller of MITRE; to Richard Katzenberg, Herman Drees and Benjamin Wolff of the American Wind Energy Association; to Benjamin Blackwell, and Jack Reed of Sandia Laboratories; to Ronald Thomas and Joseph Savino of NASA-Lewis Laboratories; and to Dorothy Berks, Reed Eldridge, and Randolph Loftus, all of whom generously contributed their time and effort.

FRANK R. ELDRIDGE

PICTURE CREDITS

Cover: Robert Berks
Graphic Arts: William Baumgardner and
Jane Andrle, MITRE/Metrek

VIABILITY

Frontispiece: NASA-Lewis

Figures 1, 2, 3, 4, 5, 6 and 7: Reproduced
from Compton's Encyclopedia, by
permission of F. E. Compton Company, a
Division of Encyclopedia
Britannica, Inc.

Figures 8 & 9: MITRE/Metrek

Figure 10: Jean Fischer, F. L. Smidth Co.,
Denmark

Figure 11: Photo Courtesy General Electric
Space Division

Figure 12: MITRE/Metrek

HISTORY

Frontispiece: Bettman Archive;
Palmer Putnam

Figure 13: Haps Wulff

Figure 14: National Geographic Society

Figure 15: Frederick Stockhuyzen

Figure 16: Frederick Stockhuyzen

Figure 17: Frederick Stockhuyzen

Figure 18: J. B. Collaert

Figure 19: Frederick Stockhuyzen

Figure 20: Frederick Stockhuyzen

Figure 21: Frederick Stockhuyzen

Figure 22: Randolph Loftus

Figure 23: MITRE/Metrek

Figure 24: Marshal Merriam

Figure 25: Central Wind Energy Institute
of Moscow

Figure 26: John Brown Co.

Figure 27: Enfield Cable Co.

Figure 28: Electricite de France

Figure 29: Ulrick Hutter

TAXONOMY

Frontispiece: Robert Plourde,
MITRE/Metrek

Figure 30: R. Powe

Figure 31: W. Linder, Saab-Scania

Figure 32: Boeing Vertol Co.

Figure 33: American Wind Turbine, Inc.

Figure 34: T. Sweeney, Princeton University

Figure 35: H. Honnef

Figure 36: Percy Thomas

Figure 37: Grumman Aerospace Corporation

Figure 38: E. Golding

Figure 39: Science Applications, Inc.

Figure 40: Science Applications, Inc.

Figure 41: R. J. Templin

Figure 42: Sandia Laboratories

Figure 43: J. Madaras

Figure 44: NASA-Lewis

Figure 45: NASA-Lewis

Figure 46: Grumman Aerospace Corporation

FUTURE POTENTIAL

Frontispiece: E. Golding; J. Fischer

Figure 47: O. Ljungstrom

Figure 48: O. Ljungstrom

Figure 49: NASA-Lewis

Figure 50: MITRE/Metrek

Figure 51: NASA-Lewis

Figure·52: NASA-Lewis

Figure 53: NASA-Lewis

Figure 54: NASA-Lewis

Figure 55: NASA-Lewis

Figure 56: MITRE/Metrek

Figure 57: General Electric Space Division

Figure 58: Boeing Engineering and
Construction Co.

Figure 59: Kaman Aerospace Co.

Figure 60: Marshal Merriam

Figure 61A: MITRE/Metrek

Figure 61B: Tvind Colleges

Figure 62: MITRE/Metrek

Figure 63: MITRE/Metrek

Figure 64: O. Ljungstrom

Figure 65: O. Ljungstrom

Figure 66: F. L. Smidth Co.

Figure 67: F. L. Smidth Co.

Figure 68: F. L. Smidth Co.

Figure 69: Grumman Aerospace Company

APPLICATIONS

Frontispiece: MITRE/Metrek

Figure 70: National Geographic Society

Figure 71: Brace Institute

Figure 72: Vickie McWhirter, MITRE/Metrek

CONTENTS

WIND MACHINES

second edition

OPERATING IN A GRID

Large wind-turbines driving electrical generators of MW-scale can be tied into public utility networks and used to save the fuel of the regular generating plants when the wind is blowing.

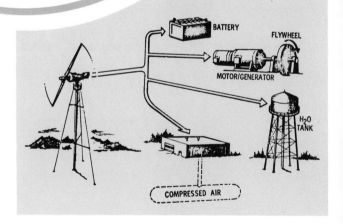

OPERATING ALONE

Wind turbines, of any size, can also be used, alone, to generate power for a variety of tasks, such as producing electricity or heat, or pumping water. However, if a continuous source of power is needed, some form of energy storage may be required, since the winds blow intermittently in most locations.

2

1. VIABILITY

WIND RESOURCES

The winds of the earth are caused primarily by unequal heating of the earth's surface by the sun. During the day, the air over oceans and lakes remains relatively cool, since much of the sun's energy is consumed in evaporating water, or is absorbed by the water itself (Figure 1). Over land, the air is heated more during the day, since land absorbs less sunlight than does water, and evaporation is less. The heated air over the land expands, becomes lighter, and rises. The cooler, heavier air from over the water moves in to replace it. In this way, local breezes on a shoreline are created.

During the night, these local seashore breezes reverse themselves, since land cools more rapidly than water, and so does the air above it (Figure 2). The warm air that rises from the surface of the water is replaced by this cool air from over the land as it blows seaward.

Similar local breezes occur on mountainsides during the day as heated air rises along the warm slopes that face the sun (Figure 3). During the night, the relatively heavy cool air on the slopes flows down into the valleys (Figure 4).

Likewise, circulating planetary winds are caused by the greater heating of the earth's surface near the equator than near the poles. This causes cold surface winds to blow from the poles to the equator to replace the hot air that rises in the tropics and moves in the upper atmosphere toward the poles (Figure 5). However, the rotation of the earth also affects these planetary winds (Figure 6). The inertia in the cold air moving near the surface toward the equator tends to twist it to the west, while the warm air moving in the upper atmosphere toward the poles tends to be turned to the east. This causes large counter-clockwise circulation of the air around low pressure areas in the northern hemisphere, and clockwise circulation around such areas in the southern hemisphere. Since the earth's axis of rotation is inclined at an angle of 23.5° to the plane in which it moves around the sun, seasonal variations in the heat received from the sun result in seasonal changes in the strength and direction of the winds at any given location on the earth's surface (Figures 7 and 8).

Sufficient energy is continually being transferred from the sun to the winds of the earth to maintain an estimated total power capacity of about 3.6×10^{15} watts in these winds.[1] This is equivalent to about 150,000 quads per year, where one quad is the total amount of energy that is avail-

3

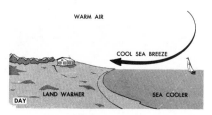

Figure 1. How land and sea breezes are caused, daytime.

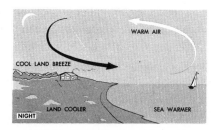

Figure 2. How land and sea breezes are caused, night time.

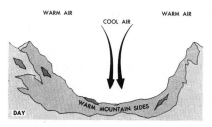

Figure 3. How mountain and valley breezes occur, daytime.

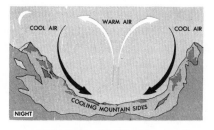

Figure 4. How mountain and valley breezes occur, night time.

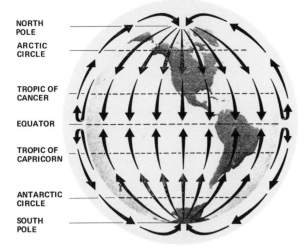

Figure 5. How winds would blow if the earth did not rotate.

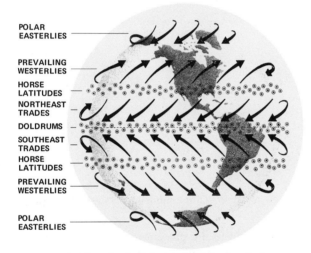

Figure 6. How the earth's rotation affects winds.

able in the coal that could be carried in 500,000 railway coal-cars, or in the oil that could be shipped in about 100 supertankers.

A recent review of the availability of wind power, by Dr. Marvin Gustavson of the Lawrence Livermore Laboratories, indicates that about 2% of the total solar flux, which averages about 350 watts per square meter of the earth's surface, is dissipated in the form of wind energy.[1] Of this, it is estimated that 35%, or about 40,000 quads per year, is extracted near the surface of the earth through surface friction and air turbulence. Assuming that as much as 10% of the near-surface wind energy could ultimately be extracted by wind machines, this results in a global extraction limit of about 4,000 quads per year, of which about 60 quads per year is estimated to be the available limit over the 48 contiguous states of the United States. This

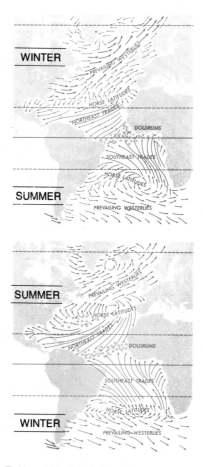

Figure 7. How the wind belts shift with the seasons.

is roughly equivalent to the present total annual energy consumption of the United States. Moreover, Dr. Gustafson estimates that the wind energy extraction limit is considerably larger than that of tidal, geothermal or hydropower (Table 1).

UNITED STATES ENERGY STATUS

In regard to present United States energy consumption, about 37% is used for residential and commercial purposes, 37% for industrial purposes, and 26% for transportation.

In 1977, sources of energy included oil (48%), natural gas (25%), coal (19%), hydropower (3%), and nuclear (4%). The remaining 1% was supplied by synthetic fuels, oil shale, geothermal energy, solar energy, wind energy, biomass, and other types of energy resources. About 26% of energy consumption in 1977 was in the form of imported oil. The present world price of oil is about $13.50 per barrel, which is equivalent to about $36 per ton of coal, or $2.33 per million Btu of oil, or $2.33 per 1,000 cubic feet of natural gas (Figure 8).

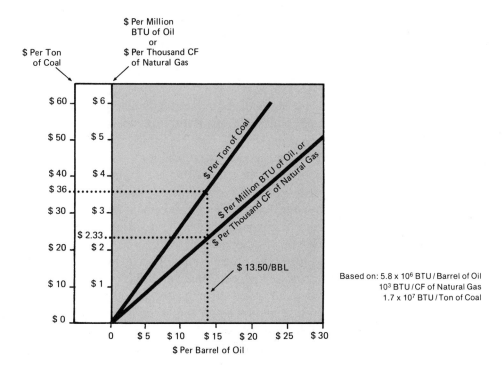

Figure 8. Equivalent costs of fuels.

Table 1. Comparison of power demands and extraction limits.
(Quads per year)

GLOBAL	
Power demand—1980	300
Extraction limits	
• Wind power	4,000
• Hydropower	120
• Geothermal power	6
• Tidal power	2
UNITED STATES (48 continguous)	
Power demand—1980	60
Extraction limits	
• Wind power	60

Although domestic oil and natural gas production is expected to decrease rapidly in the next few decades, total United States energy consumption is expected to increase to about 115 quads per year by the year 2000, unless stringent conservation measures are applied.

FEDERAL PROGRAM

Wind energy supplied a significant amount of the energy consumed in rural areas of the United States until the Rural Electrification Administration (REA) introduced electrical cooperatives in the mid-1930's. However, with the emergence of the energy crunch in the early 1970's, there has been a renewed interest in the use of wind energy for these and other applications.

As a result of this interest, the federal government has undertaken an accelerated Wind Energy Conversion System (WECS) program[2] with the objective of stimulating the development of WECS capable of handling a significant amount of the United States' energy needs by the year 2000. To accomplish this, a series of experimental WECS units, from one kilowatt to several megawatts rated capacity, are being developed. These developments are being supported by an extensive research and technology effort, which will lead to the testing, evaluation, and demonstration of units of all sizes by the early 1980s, including 10- to 100-megawatt multi-unit demonstration systems, jointly funded with federally-owned utilities and other users. While the emphasis in this federal program will be placed on the use of wind machines to produce electricity, a number of other applications, such as fuel generation, the production of direct heat for heating water and air-space, crop drying, and fertilizer manufacturing, will also be demonstrated, particularly through the use of the smaller sizes of wind machines.

There are a number of possible constraints to the rate at which WEC systems can be expected to be developed and applied. These are being intensively addressed in the federal wind energy research and technology program. They include uncertainties in:

- The availability of wind energy at specific potential installation sites;
- Capital, operations, and maintenance costs;
- The economic viability of various types of WECS applications;
- The availability of capital to build WEC systems;
- The environmental and public acceptability of such systems; and
- Various legal and institutional problems that may arise from the use of WEC systems.

ECONOMICS

An important measure of the economic viability of WECS is the cost of electricity produced by such systems. This cost is still relatively large

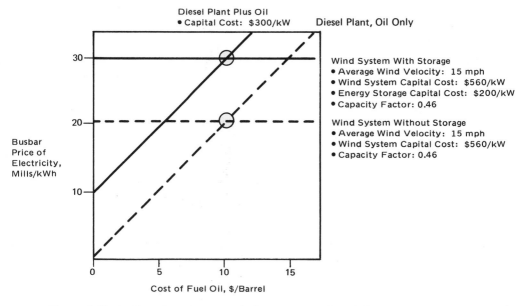

Figure 9. Typical break-even costs: wind systems vs. diesel plants.

compared to that of electricity produced by nuclear or fossil-fueled systems. However, the unit price of wind turbines is expected to decrease as mass-production techniques are introduced and mass-distribution and installation procedures are used. Current estimates indicate that, ultimately, both large- and small-scale WECS, if located at typical sites with average wind velocities of about 15 mph, would be expected to produce electricity at busbar prices of 20 to 30 mills per kilowatt-hour.

Wind systems without energy storage may, for instance, be tied into a conventional system, powered by a diesel-electric generating plant and used to save fuel when the wind is blowing. On the other hand, wind systems with suitable energy storage capabilities can be used as a complete substitute for a conventional electric generating plant. In either case, the break-even points might be expected to occur at fuel prices of about $10 to $11 per barrel of oil for sites with average wind velocities of about 15 mph at hub height (Figure 9), or about $13.50 per barrel for velocities of 14 mph at hub height. The latter wind speed is available, at 50 meters, above approximately 35% of the land area of the United States. To achieve these break-even prices at even more moderate wind velocities will require the development of systems with either improved performance or lower life-cycle costs.

PUBLIC ACCEPTANCE

Public acceptance of wind energy conversion systems is an important consideration in planning for the widespread application of wind energy. Preliminary studies have shown that the environmental impact of such

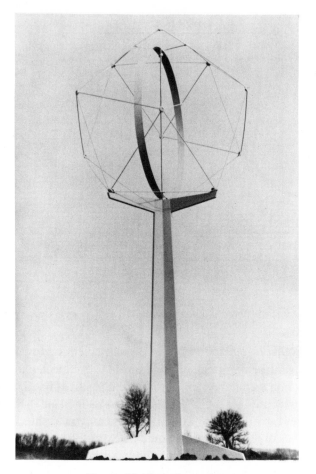

Figure 10. Vertical-axis rotor.

systems is relatively small compared to conventional power systems.[3] Wind-powered systems do not require the flooding of large land areas or the alteration of the natural ecology, as hydroelectric systems do. Furthermore, they produce no waste products or thermal or chemical effluents, as fossil-fueled and nuclear-fueled systems do.

Conventional wind turbine systems that generate several megawatts of power require large exposed rotors several hundred feet in diameter, located on high towers (Figures 10 and 11). The rotors of such systems, being passive, are practically noiseless. However, special precautions may be necessary to prevent them from causing interference with TV picture signals or AM radio at nearby receivers,[4] and some safety measures may be required to prevent damage or injury from possible mishaps in cases where there is danger that the rotors might break or shed ice.[5]

The only other concerns with conventional wind machines are those of aesthetics. Large numbers of units and interconnecting transmission lines will be required in the future if such systems are to have any significant

Figure 11. Horizontal-axis rotor.

Figure 12. Building with confined wind vortex generator.

impact on the United States' energy demands. Particular attention is being given, therefore, to the development of attractive designs for the towers, rotors, and nacelles of such systems to avoid "visual pollution," and new designs, such as vortex generators, are being developed that could be incorporated into buildings and other structures (Figure 12). Many of these new types of systems would use high-speed, enclosed rotors that would not cause television or AM radio interference and might be utilized effectively even in cities and other urban areas where conventional types of wind rotors might be banned because of aesthetic, interference, or safety reasons.

SAILING ON THE NILE

This 5000-year-old drawing of a Nile river craft is the oldest recorded depiction of the use of a sail to produce translational motion. · Sails were later mounted on booms attached to a central spindle to drive rotational devices, such as wheels for pumping irrigation water and mills for grinding grain or sawing wood.

GENERATING ELECTRICITY IN VERMONT

The largest wind turbine ever built was located on top of 2000 foot Grandpa's Knob near Rutland, Vermont. Mounted on its 110 foot tower, and having rotor blades measuring 175 feet from tip to tip, it generated 1.25 megawatts of electrical power in winds of 30 miles per hour or more. Conceived by Palmer Putnam and funded by the S. Morgan Smith Company, it ran intermittently from October 1941 to March 1945, when it was shut down after a blade failure. Actually, this technical development problem could have been overcome but the economics of that time did not permit the system to compete with cheap oil and coal.

2. HISTORY

ANCIENT HISTORY

The first wind turbines were probably simple vertical-axis panemones, such as those used in Persia as early as about 200 B.C. for grinding grain (Figure 13). The use of these vertical-axis mills subsequently spread throughout the Islamic world.

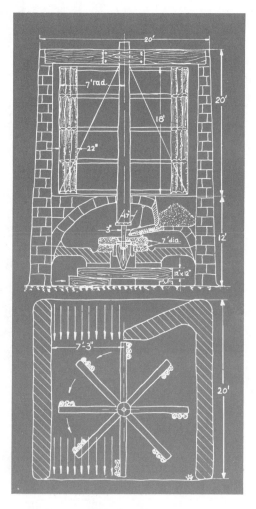

Figure 13. Persian vertical-axis windmill with grindstone, in sectional views. This version had bundles of reeds instead of cloth for its sails.

Later, horizontal-axis windmills, consisting of up to ten wooden booms, rigged with jib sails, were developed. Such primitive types of windmills are still found in use today in many Mediterranean regions (Figure 14).

Figure 14. Primitive type of horizontal-axis windmill with jib sails wrapped on wooden booms. This windmill is still being used today on the Greek island of Mikonos for grinding grain.

MIDDLE AGES

By the eleventh century A.D., windmills were in extensive use in the Middle East and were introduced to Europe in the thirteenth century by returning Crusaders.

During the Middle Ages in Europe, most manorial rights included the right to refuse permission to build windmills, thus compelling tenants to have their grain ground at the mill of the lord of the manor. Planting of trees near windmills was banned to ensure "free wind."

By the fourteenth century, the Dutch had taken the lead in improving the design of windmills and used them extensively thereafter for draining the marshes and lakes of the Rhine River delta (Figures 15 and 16). Be-

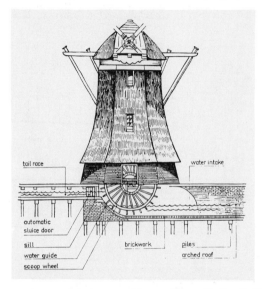

Figure 15. Large octagonal drainage mill or polder mill, South Holland type, with internal scoop wheel. The tower in front elevation, the lower part in section, to show the working of the scoop wheel.

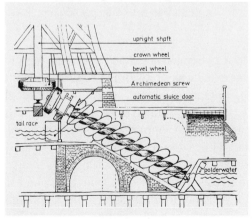

Figure 16. Arrangement showing method of driving the Archimedean screw of a drainage mill.

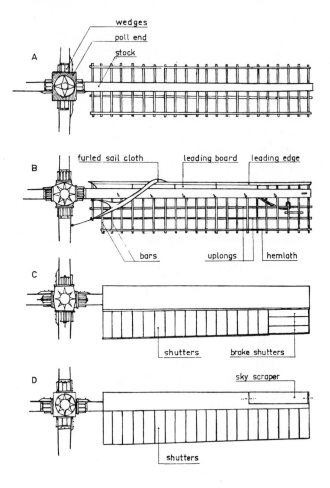

Figure 17. Types of sail:
A. Oldest type, double-sided (about 1600)
B. Normal old-fashioned Dutch type (one leading board taken away)
C. Shuttered type with air brake
D. Shuttered type with sky scraper.

tween 1608 and 1612, Beemster Polder, a wetland area which was about 10 feet below sea level, was drained by 26 windmills of up to 50 horsepower (hp) each, operating in two stages. Later, Leeghwater, the renowned hydraulic engineer, drained Schermer Polder in four years. Fourteen windmills pumped water into a storage basin at a rate of 1,000 cubic meters per minute. Thirty-six mills then pumped the storage basin water into a ring canal that emptied into the North Sea.

The first oil mill was built in Holland in 1582, and in 1586 the first paper mill was constructed, to meet the enormous demands for paper that resulted from the invention of the printing press. At the end of the sixteenth century, sawmills were introduced to process timber imported from

Figure 18. Engraving by J. B. Collaert (1566-1628) showing windmills of that era with sail bars on both sides of the stocks.

the Baltic regions. By the middle of the nineteenth century, some 9,000 windmills were being used in the Netherlands for a wide variety of purposes.

With the introduction of the steam engine, during the Industrial Revolution, the use of windpower in Holland started to decline, and by the turn of the twentieth century, only about 2,500 windmills were still in operation in the Netherlands. By 1960, fewer than 1,000 were still in working condition.

The Dutch introduced many improvements in the design of windmills and, in particular, the rotors. By the sixteenth century, the primitive jib sails on wooden booms had given way to sails supported by wooden bars on both sides of the stock (Figures 17 and 18). Later, the bars were moved to the trailing edge of the rotor to improve the aerodynamic design (Figure 19). More modern designs substituted sheet metal for the cloth sails, used steel stocks, and introduced various types of shutters and flaps to control the speed of the rotor in heavy weather. Large industrial mills could deliver up to 90 hp in high winds (Figures 20 and 21).

RECENT EXPERIENCE

United States Experience

Since the mid-nineteenth century, more than six-million small multi-bladed windmills, providing power outputs of less than 1 hp each in an average wind, have been built and used in the United States to pump wa-

Figure 19. Drainage mill, North-Holland type, built in 1761. The cap is turned from the inside, hence no tail pole is necessary. This is one of the mills of the Zype and Haze polders in the province of North Holland.

Figure 20. Wooden gear-train for typical Dutch windmill.

Figure 21. Interior view of Dutch windmill used for grinding grain.

Figure 22. Pumping water.

ter, generate electricity, and perform similar functions. It is estimated that over 150,000 are currently in operation.

Water pumping windmills are used in many parts of the United States, not only for pumping water for farm and rural households, but for watering livestock on ranges in remote areas. These types of machines commonly have metal fan-blades, 12 to 16 feet in diameter, mounted on a horizontal shaft, with a tail-vane to keep the rotor facing into the wind (Figure 22). The shaft is usually connected to a set of gears and a cam that move a connecting rod up and down. This rod, in turn, operates a pump at the bottom of the tower. A 12-foot diameter rotor of this type develops about 2/3 hp in a 15 mph wind and can pump about 10 gallons of water per minute to a height of about 100 feet.

Small wind machines, used to generate electricity, usually have two or three propeller-type blades that are connected by a shaft and gear train to a d.c. generator (Figure 23). They usually incorporate some type of energy storage system, often consisting of a bank of batteries.

Figure 23. Generating electricity.

One of the classic designs of this type is the Jacobs Wind Electric Company unit with a three-blade propeller, 14 feet in swept diameter. This type of unit can deliver about 1 kilowatt (kW) in a wind of 14 mph, which is about average for the United States at about 20 meters above ground level. A typical American household, without electrical heating of air-space, uses an average of about 1 kW of electrical power, so a single unit of this type could be used to supply the electrical needs of an average household under these conditions.

Between 1930 and 1960, tens of thousands of these Jacobs wind-powered electric generators were sold and installed in many countries throughout the world. Production stopped in 1960, after the Rural Electrification Administration succeeded in supplying most American farms and rural homes with inexpensive electricity through cooperative utilities.

The Jacobs design uses a rotational-speed-actuated governor for varying the pitch of the propeller blades. This method has proved very successful in protecting these machines against high winds, including hurricanes in Florida, the Gulf Coast, and throughout the West Indies, as well as against storms in the Arctic Region and Antarctica. For instance, a Jacobs machine, installed on top of one of the original 70 foot radio towers at Little America, Antarctica, in 1933, was reported by Admiral Byrd's son to be still in excellent operating condition in 1955, although the snow by that time had drifted to within only 10 feet of the top of the tower.

Records from several thousand Jacobs wind-powered generators installed throughout the United States during the period from 1930 to 1960 showed that maintenance costs for these machines averaged less than $10 per year (in 1960 dollars). Since the recent oil crunch, many of the original Jacobs machines have been refurbished and are being put back into service.

Another interesting wind machine that was built during that period was the Smith-Putnam unit.[6] This was the largest operational wind-powered electric generator that had been built until recently. After a lengthy study in the 1930's, Palmer Putnam had concluded that a large machine was required to minimize the cost of electricity generated by the wind. With the assistance of the eminent Cal Tech aerodynamicist Theodore Von Karman, and various members of the MIT staff, he designed a large wind turbine to feed power into the existing electrical network of the Central Vermont Public Service Company. The S. Morgan Smith Company of York, Pennsylvania, constructed and operated the plant in the early 1940's. The two-bladed, 175 foot diameter, propeller-type rotor weighed 16 tons and operated at a constant rotational speed of 28 rpm to produce up to 1.25 megawatts (MW) of a.c. power.

In March 1945, after intermittent operation over a period of several years, one of the blades broke off near the hub, where a known weakness had previously been identified but had not been corrected because of war-time material shortages. A comprehensive economic study indicated that the plant, even if repaired, could not compete effectively at that time, with conventional electrical generation plants, so the project was abandoned.

Danish Experience

In Denmark, by the end of the nineteenth century, there were about 2,500 industrial windmills in operation, supplying a total of about 40,000 hp or 30 MW; i.e., about 25% of the total power available to Danish industry at that time.[7] About 94% of these windmills were in rural areas and only about 6% in towns. In addition, approximately 4,600 windmills were also being used on about 2% of the Danish farms for various applications, including threshing, milling of grain, and water pumping. Other wind machines were used for pumping water for livestock and for draining wetlands. Most of these machines were located in the western and northern parts of Jutland, where the winds are strongest.

By the 1930's, the number of industrial windmills had declined to about 1,000, but the number of farm units had increased to about 16,000. By the end of World War II, about 1,500 farms had small-scale wind machines that were used for the production of electricity. All of these machines went out of service rapidly after World War II, as cheap oil and gas became available in Denmark.

Meanwhile, in 1890, the Danish government had initiated a program to develop large-scale wind-powered electric generators. By 1908, 72 systems designed by Professor P. La Cour had been built. These machines, with rated output capacities in the range of 5 to 25 kW, consisted of 80 foot towers supporting a 75 foot diameter, four-bladed rotor that operated a mechanical gear-and-shaft drive-train connected to an electrical generator on the ground. The number of such wind machines reached 120 by 1918, but decreased after World War I.

During World War II, the Danes developed and operated a number of new types of large-scale wind machines for producing electricity. The number of these machines increased from 16 in the summer of 1940, to 88 by the beginning of 1944. After World War II, the number of operating machines started to decrease, and at the end of 1947, they dropped to 57. The decrease continued in the 1950's and by the end of that decade the production of electricity by wind machines was being conducted only on an experimental basis in Denmark, with units rated at 12, 45, and 200 kW, which were tied into existing public utility networks. The 200 kW Gedser mill (Figure 24), which was the latest in this series, was operated until 1968, when it was shut down because it was found that by that time the cost of electricity supplied by this wind-powered unit was about twice

Figure 24. Restored Danish Gedser wind-turbine; 200-kW rated output power at 33.6 mph rated wind speed.

the equivalent fuel cost of the steam-powered electric utility plants that were being operated in Denmark. After the energy crunch of 1973, the 200 kW Gedser mill was refurbished, and in 1977 it was put back into service, using funding partly supplied by the United States Department of Energy.

Russian Experience

In 1931, the Russians built an advanced 100 kW wind turbine (Figure 25) near Yalta on the Black Sea. The turbine was tied, by a 6,300 volt line, to a 20 MW steam-powered station 20 miles away. The annual output of this machine was found to be about 280,000 kWh per year, which yielded a capacity factor of 0.32. The generator and controls were located on top of a 100 foot tower. The speed of the rotor was regulated by controlling the pitch of the blades. The heel of the inclined strut was mounted on a car-

Figure 25. Russian wind-turbine; 100 kW (24.6 mph rated wind speed).

riage that ran on a circular track to keep the rotor facing into the wind. At one time in the 1930's, the Russians considered building a larger system of 5 MW rated capacity, but apparently this project was never implemented.

British Experience

In England, in the late 1940's and during the 1950's, considerable work was done on wind-powered electrical generation plants under the leadership of E. Golding and A. Stoddard.[8] Wind measurements made at about 100 sites in the British Isles during that period are among the best documented results available on wind characteristics in a given geographical region. In 1950, the North Scotland Hydroelectric Board commissioned the John Brown Company to construct an experimental wind turbine on Cape Costa in the Orkney Islands (Figure 26). This unit was designed to generate 100 kW in winds of 35 mph. It operated for short periods in 1955, coupled to a diesel-powered electric utility network, but was shut down because of operational problems.

In the 1950's, the Enfield Cable Company built a unique 100 kW wind-powered generator (Figure 27) designed by a Frenchman named Andreau. This machine was operated at St. Albans, England, and later in

Figure 26. British wind-turbine, John Brown design located on Cape Costa in the Orkney Islands; 100 kW (35 mph rated wind speed).

Algeria. It consisted of a hollow tower, 85 feet high, and a hollow rotor 80 feet in diameter, with openings at the blade tips. This caused a pressure differential that drove the air from openings near the base of the tower, up the tower, through an air turbine in the tower, and out of the rotor blade tips. The efficiency of the unit was found to be low compared to more conventional horizontal-axis wind-powered rotors.

French Experience

The French built and operated several large wind-powered electric generators in the period from 1958 to 1966.[9] These included three horizontal-axis units, each with three propeller-type blades.

A unit of this kind was operated intermittently near Paris from 1958 to 1963 (Figure 28). This unit was designed to generate 800 kW in winds of 37 mph. The rotor (100 foot swept diameter), generator, and gear assembly weighed 160 tons and was mounted on a 100 foot tower. The rotor was operated at a constant speed of 47 rpm and was coupled to a 50 Hz,

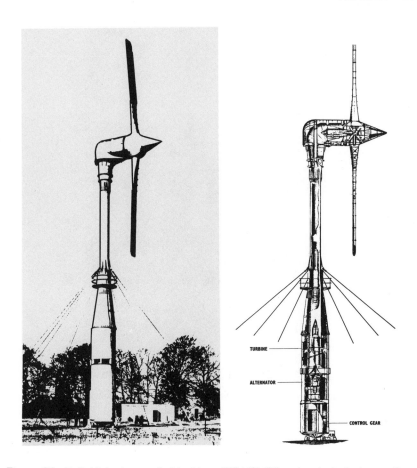

Figure 27. Enfield-Andreau wind-turbine; 100 kW (33 mph rated wind speed).

60 kV utility grid, using a synchronous alternator, operating at 1,000 rpm and generating current at 3,000 volts. This was increased to 60,000 volts by voltage transformers that were connected to the main grid by a 15 kilometer transmission line.

Two other units were constructed at St. Remy-des-Landes in Southern France. The smaller unit had a 70 foot diameter rotor operated at 56 rpm. An asynchronous generator with a nominal speed of 1,530 rpm generated 132 kW in winds of 28 mph or more. The larger of these two units was rated at 1,000 kW in winds of 37 mph and weighed 96 tons, excluding the tower.

The capital cost of the Paris unit was about $1,155/kW, while that of the St. Remy units was about $1,000/kW (in 1960 dollars).

The French also built and tested several experimental vertical-axis panemones during this period.

Figure 28. French wind-turbine located near Paris, France; 800 kW
(36 mph rated wind speed).

Figure 29. German wind-turbine, Hutter-Allgaier design; 100 kW
(18 mph rated wind speed).

German Experience

The Germans, under the direction of Professor Ulrich Hutter, introduced a number of improvements in the design of wind-powered generators, including light-weight constant-speed rotors that were controlled by variable-pitch propeller blades with swept diameters as large as 110 feet.[10] These machines used light-weight, composite carbon-epoxy, or composite fiberglass-epoxy blades, with the generator mounted on a tower consisting of a small-diameter hollow pipe, supported by guy wires (Figure 29). The largest unit generated 100 kW in 18 mph winds. These units operated successfully for more that 4,000 hours during the period from 1957 to 1968. This work resulted in some of the most advanced wind turbines that have yet been built. The light-weight fiberglass blades resulted in fewer bearing and blade failures than were experienced with the heavier machines built in other countries. The wind machine with the 110 foot swept diameter rotor was in operation for many years. The blades were found to be sound and undamaged when they were finally dismantled and examined.

TAXONOMY

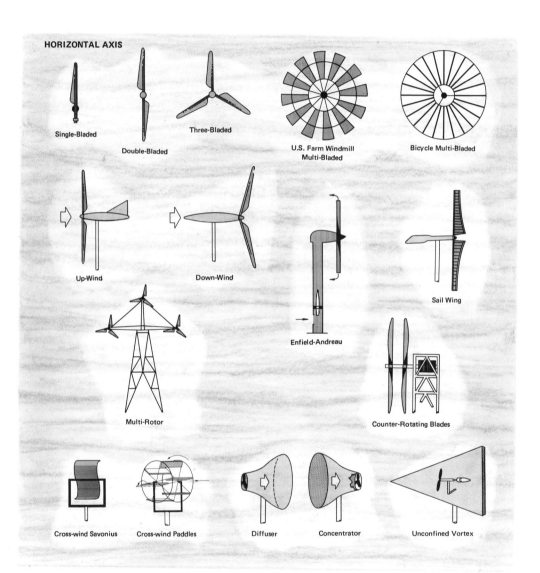

HORIZONTAL AXIS

Single-Bladed

Double-Bladed

Three-Bladed

U.S. Farm Windmill
Multi-Bladed

Bicycle Multi-Bladed

Up-Wind

Down-Wind

Sail Wing

Enfield-Andreau

Multi-Rotor

Counter-Rotating Blades

Cross-wind Savonius

Cross-wind Paddles

Diffuser

Concentrator

Unconfined Vortex

VERTICAL AXIS

PRIMARILY DRAG-TYPE

Savonius

Multi-Bladed
Savonius

Shield

Plates

Cupped

PRIMARILY LIFT-TYPE

φ-Darrieus

Δ-Darrieus

Giromill

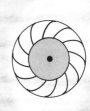

Turbine

COMBINATIONS

Savonius/φ-Darrieus

Split Savonius

Magnus

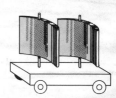

Airfoil

OTHERS

Deflector

Sunlight

Venturi

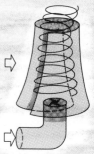

Confined Vortex

3. TAXONOMY

TYPES OF WIND ENERGY COLLECTORS

Many types of wind energy collectors have been devised. It is said that more patents for wind systems have been applied for than for nearly any other type of device. Basically, almost any physical configuration which produces an asymmetrical force in a windstream can be made to rotate, translate, or oscillate, and power can be extracted. The important question of course is: How much power for how much size and cost?

Machines using rotors as wind energy collectors may be classified in terms of the orientation of their axis of rotation, relative to the windstream.

- *Horizontal-axis rotors (head-on)*—for which the axis of rotation is parallel to the direction of the windstream; typical of conventional windmills.
- *Crosswind horizontal-axis rotors*—for which the axis of rotation is both horizontal to the surface of the earth and perpendicular to the direction of the windstream; somewhat like a water wheel.
- *Vertical-axis rotors*—for which the axis of rotation is perpendicular to both the surface of the earth and the windstream.

In addition, a number of types of translational wind machines have been devised, including the sailing ship itself, sailing ships that carry water-driven turbines mechanically connected to an electric generator, and land vehicles driven by sails or solid airfoils on a closed track or roadway (Figure 30), with their wheels mechanically linked to an electric generator. Other types of translational devices have been designed to produce power by vibrating or oscillating in a windstream.

Still other types of wind energy conversion devices are being developed that use no moving parts, such as devices that use differential cooling in a windstream to generate electricity by means of the Thomson thermoelectric effect, or devices that use wind energy to drive charged aerosol particles toward a collector electrode.

HORIZONTAL-AXIS ROTORS

Head-On

Horizontal-axis rotors can be either lift or drag devices. Lift devices (Figure 31) are generally preferred, since for a given swept area, high rota-

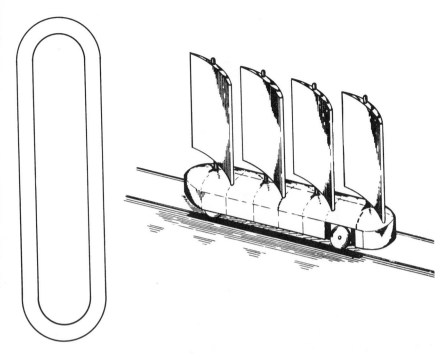

Figure 30. Example of a solid-airfoil tracked-vehicle used for generating electricity.

tional speeds and more output power can be developed by lift than by drag forces. In general, a drag device cannot move faster than the wind velocity, while a lift device can. Thus a lifting surface can obtain higher tip-to-wind speeds and, consequently, a higher power output to weight ratio and, for many applications, a lower cost to power output ratio.

Either lift or drag-type rotors can be designed with different numbers of blades, ranging from one-bladed devices with a counterweight (Figure 32), to devices with large numbers of blades, i.e., up to 50 or more (Figure 33).

The ratio of the projected area of a rotor (on a plane perpendicular to its axis of rotation) to the swept area of the rotor is known as the "solidity" of the rotor. For multi-vane fan-type rotors such as the American farm windmill, a typical solidity is 0.7. For high-speed lift-type propellers, on the other hand, the solidity is usually much lower, i.e., 0.1 to 0.01.

Lift-type rotors often use tapered and/or twisted blades to reduce the bending strains on the roots of the blades.

Some horizontal-axis rotors are designed to be yaw-fixed; i.e., they cannot be rotated around a vertical axis perpendicular to the windstream. Generally, this type would only be used where there are prevailing winds from one direction. Most types are yaw-active and will "track" the changing direction of the wind. Many small-scale systems have been designed to yaw using a tail-vane, whereas larger systems are often designed as down-wind rotors and may use canted blades to help orient them

Figure 31. Downwind, lift-type, horizontal-axis wind machine built by Saab-Scania of Sweden, 60kW (25 mph rated wind speed).

in the windstream. Large-scale systems with downwind rotors are usually servo operated to orient them properly in yaw.

A number of means are used to prevent a rotor from over-speeding in high winds, including feathering of the blades, using flaps that rotate with the blades, installing flaps on the blades themselves, or using spring-loaded mechanical devices that turn the rotor sideways to the windstream as the windspeed increases.

Some blades are directly coupled to the output of the system through a shaft on which the rotor is mounted (Figure 31). Others use a circular rim attached to the blade-tips to drive a secondary shaft that is mechanically connected to an electric generator or other form of power output (Figure 33).

Modern descendants of the ancient jib-sail rotors, designed by Princeton University and others, are manufactured from metal parts and cloth.[11]

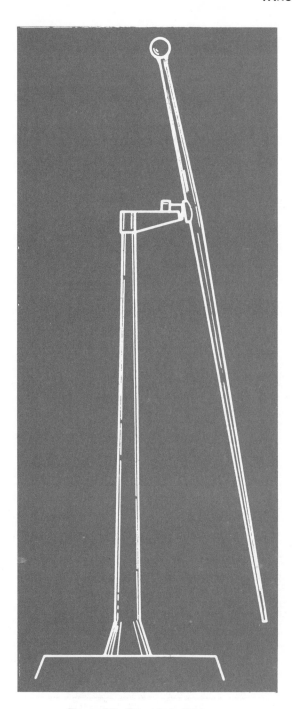

Figure 32. One-bladed rotor.

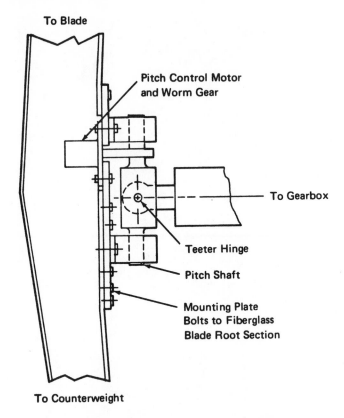

To Blade

Pitch Control Motor and Worm Gear

To Gearbox

Teeter Hinge

Pitch Shaft

Mounting Plate Bolts to Fiberglass Blade Root Section

To Counterweight

BLADE CONTROLS FOR ONE-BLADED ROTOR

ONE-BLADED ROTORS	
ADVANTAGES	**DISADVANTAGES**

• SIMPLER BLADE CONTROLS
  — LOWER BLADE WEIGHT AND COST
  — LOWER GEAR BOX COST
	• VIBRATION PRODUCED, DUE TO AERODYNAMIC TORQUE.

• COUNTERWEIGHT COSTS LESS THAN A SECOND BLADE	• UNCONVENTIONAL APPEARANCE
• COUNTERWEIGHT CAN BE INCLINED TO REDUCE BLADE CONING	• LARGE BLADE-ROOT BENDING MOMENT
• PITCH BEARINGS DO NOT CARRY CENTRIFUGAL FORCE	• STARTING-TORQUE REDUCED BY GROUND BOUNDARY LAYER
• BLADE ROOT SPAR CAN BE LARGE DIAMETER, i.e. MORE RUGGED	• ONE-PER-REV CORIOLIS TORQUE PRODUCED, DUE TO FLAPPING

Figure 33. Multi-bladed horizontal-axis rotor.

These have the general shape of solid propellers and are known as "sail-wings." A typical sailwing (Figure 34) consists of a rigid tubular leading-edge to which short bars are attached to form a rigid tip and root. A cable is stretched between the free ends of the tip and root bars to serve as a trailing edge for the blade. The blade surface consists of a sleeve óf dacron sailcloth that is slipped over the tubular leading edge and the cable and stretched by tightening the cable. Sailwings such as this have been designed that have a lift to drag ratio comparable to conventional rigid propeller blades, but can be as much as 50% lighter in weight than their rigid counterparts. The sizes of sailwing rotors are usually limited to about 30 feet in diameter by the strengths of the materials used.

Sailwing rotor

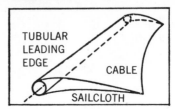

Figure 34. Sailwing rotor.

Various types of systems have also been developed with counter-rotating blades on the same axis. Others have been proposed (Figures 35 and 36) with multiple rotors on a single tower to reduce tower costs for a given power output of the system. Others (Figure 37) use tapered shrouds to concentrate and/or diffuse the windstream as it passes a rotating turbine, thereby increasing the windstream velocity and reducing turbulence. Still others are designed to produce vortices on the downstream side of a turbine to increase the pressure drop across the turbine and increase the speed of the windstream through the turbine.[12]

Crosswind

Various horizontal crosswind devices have been developed but have not been found to be very effective, since they must be turned into the wind as the wind direction changes, just as must be done with conventional head-on horizontal-axis rotors. As a result, relatively complicated means

Figure 35. Design of a multi-rotor wind machine proposed by Honnef of Berlin in 1933. This unit was to stand 1000 feet high and the inventor rated it at 50 MW.

must be used to collect the output power from such devices, which generally results in a loss of efficiency of the system. As a whole, there appear to be no significant advantages of crosswind horizontal-axis rotors over either head-on horizontal-axis rotors or vertical-axis rotors.

VERTICAL-AXIS ROTORS

In general, vertical-axis rotors have a major advantage over horizontal-axis rotors. They do not have to be turned into the wind as the direction of the windstream varies. This reduces the design complexity of the system and decreases gyro forces on the rotors, when yawing, that stress the blades, bearings, shafts, towers, and other components that are used in horizontal-axis rotor systems.

Various types of vertical-axis panemones (Figure 38) have been developed in the past that use drag forces to turn rotors of different shapes.

Figure 36. As conceived by Percy Thomas, this multi-rotor wind machine would be mounted on a 475-foot tower and would generate 6.5 MW at a rated wind speed of 28 mph.

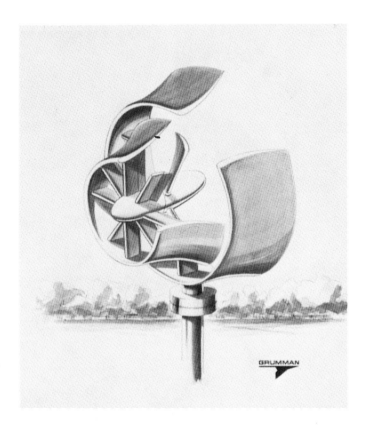

Figure 37. Diffuser-augmented turbine.

These include those panemones that use plates, cups, or turbines as the drag device, as well as the Savonius S-shaped cross-section rotors which actually provide some lift force but are still predominantly drag devices (Figures 39 and 40). Such devices have relatively high starting torques compared to lift devices because of their higher solidity, but have relatively low tip-to-wind speeds and lower power outputs per given rotor size, weight, and cost.

The Darrieus-type rotor (Figures 41 and 42) was invented by G.J.M. Darrieus of France in the 1920's and has been under extensive development by the National Research Council of Canada since the early 1970's, and by the Sandia Laboratories of the United States since 1974.[13] This

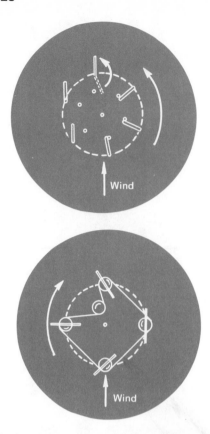

Figure 38. Articulated panemones with moveable paddles or airfoils.

CHARACTERISTICS:
- Self-Starting
- Low Speed
- Low Efficiency

Figure 39. Savonius rotor.

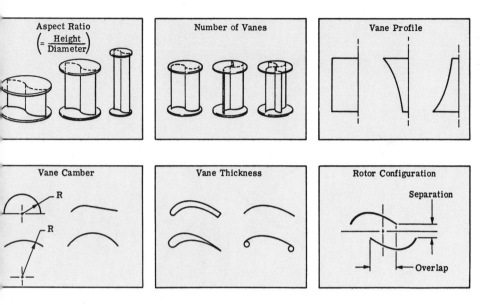

Figure 40. Savonius rotor design alternatives.

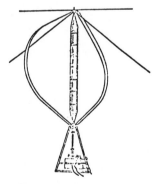

CHARACTERISTICS:
- Not Self-Starting
- High Speed
- High Efficiency
- Potentially Low Capital Cost

Figure 41. Darrieus rotor.

Figure 42. Darrieus-type vertical-axis rotor built by Sandia Laboratories, Albuquerque, New Mexico.

design is now considered a potential major competitor to the horizontal-axis propeller-type systems.

Darrieus-type rotors are lift devices, characterized by curved blades with airfoil cross-sections. They have relatively low solidity and low starting torques, but high tip-to-wind speeds and, therefore, relatively high power outputs per given rotor weight and cost. Various types of Darrieus rotor configurations have been conceived, including the ϕ-Darrieus, the Δ-Darrieus, the Y-Darrieus, and the $\diamond$-Darrieus. Such Darrieus rotors

Figure 43. Test of a pilot model of a Madaras rotor at Burlington, New Jersey, October 1933.

can be designed to operate with one, two, three, four, or more blades.

Darrieus rotors can also be combined with various types of auxiliary rotors to increase their starting torques. However, such additions increase the weight and cost of the system, so trade-offs in these characteristics must be considered in developing an optimum design for a given application.

Other types of vertical-axis rotors include Magnus Effect rotors, typified by the "Madaras" or "Flettner" designs, which consist of spinning cylinders (Figure 43). When spun in a windstream, translational forces are produced perpendicular to the windstream by the Magnus Effect (Fig-

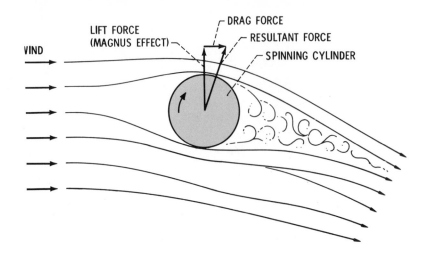

Figure 44.   The Magnus effect.

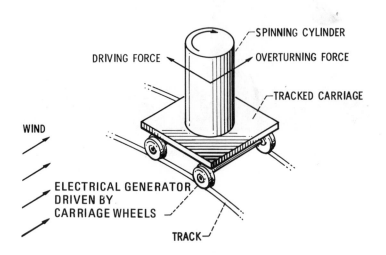

Figure 45.   The Madaras concept for generating electricity.

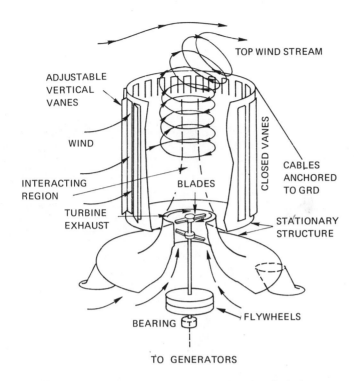

TOP WIND STREAM

ADJUSTABLE
VERTICAL
VANES

CLOSED VANES

WIND

CABLES
ANCHORED
TO GRD

INTERACTING
REGION

BLADES

TURBINE
EXHAUST

STATIONARY
STRUCTURE

FLYWHEELS

BEARING

TO GENERATORS

Figure 46. Vortex tower for omnidirectional winds.

ure 44). Such a device can be used as a sail to propel ships or land vehicles (Figure 45).

Still other types of vertical-axis rotors have been conceived that use ducts and/or vortex generator towers[12] (Figure 46), augmented by shrouds or diffusers to deflect the horizontal windstream to a vertical direction and increase its velocity, sometimes with the addition of heat, either by direct solar insolation or by burning a fuel of some type, in the final extension. The latter becomes, in effect, a type of gas turbine.

While many other types of configurations have been conceived, the primary consideration in designing a wind-powered rotor for a given application is the amount of energy per unit system cost that is derived at a given wind speed.

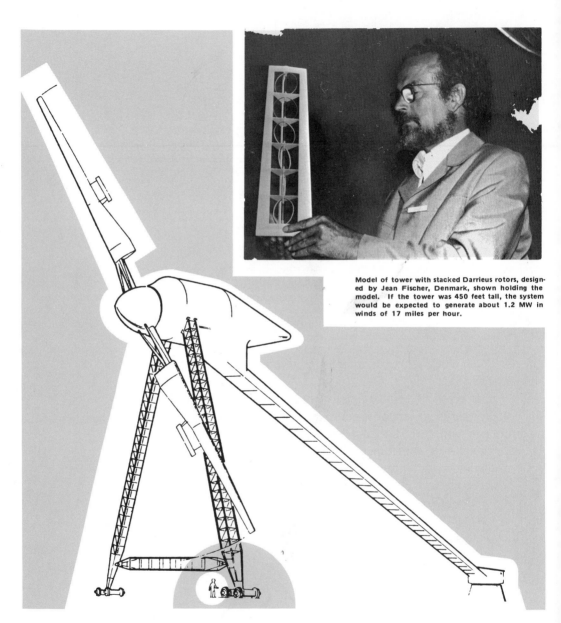

Model of tower with stacked Darrieus rotors, designed by Jean Fischer, Denmark, shown holding the model. If the tower was 450 feet tall, the system would be expected to generate about 1.2 MW in winds of 17 miles per hour.

Conceptual drawing of a 1500 KW wind generator, based on a 1950 British Design (Golding 1954)

50

4. FUTURE POTENTIAL

PROGRAM GOALS

There is an urgent need to develop alternative clean, competitive sources of energy as soon as possible. Sharp increases have occurred recently in the price of energy derived from conventional sources and systems. The rapidly growing dependence of the United States on imports of foreign oil and the extended public awareness of the potential impact of conventional energy systems on the environment have created an intensified demand for viable alternative energy sources. The use of wind energy conversion systems is one such alternative.

FEDERAL SUPPORT

Federal support is being provided for the WECS program to accelerate the design, construction, and operation of a wide range of sizes of wind machines (Figures 47 and 48), including the smaller units of 1 kW to 100 kW that are needed for many types of dispersed applications, as well as the larger types of WECS units (1 MW or more rated capacity) needed to tie in economically with public utility networks. Applications of this entire range of sizes of wind machines appear increasingly competitive as the prices of conventional fuel and electricity rise. However, federal incentives may be needed to spur these applications in the near future.

The key hurdles to be cleared in these applications are the demonstration of reliable wind machines with expected lifetimes of 20 to 30 years and a reduction in the capital costs of these units to a level of $300 to $1,000 per rated kW through inproved engineering designs and through mass-production, mass-distribution, and mass-installation of the machines.

Other important problems that are foreseen involve transferring the federally developed WECS technology to industry, public utilities, farm operators, home owners, and other users, and obtaining the necessary capital required to build facilities, for manufacturing wind machines of different sizes, rapidly enough to meet the program goals.

The planned WECS program involves a significant amount of research and development on such subjects as WECS mission and regional analyses; wind characteristics; advanced system concepts, components, and subsystems; and possible environmental, societal, legal, or institutional problems that might constrain rapid implementation.

51

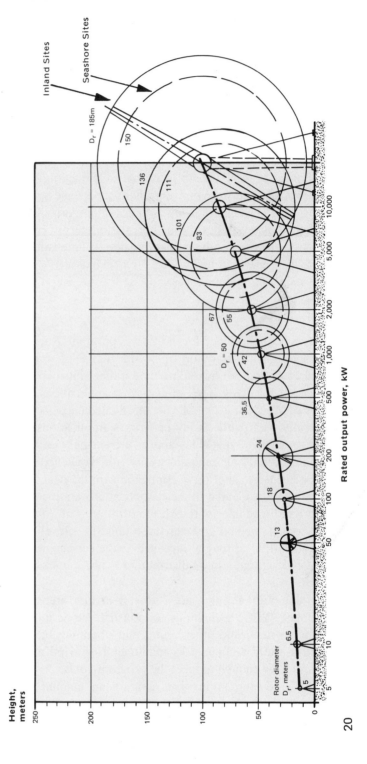

Figure 47. Typical family of horizontal-axis wind turbines for average wind speeds of 17 mph.

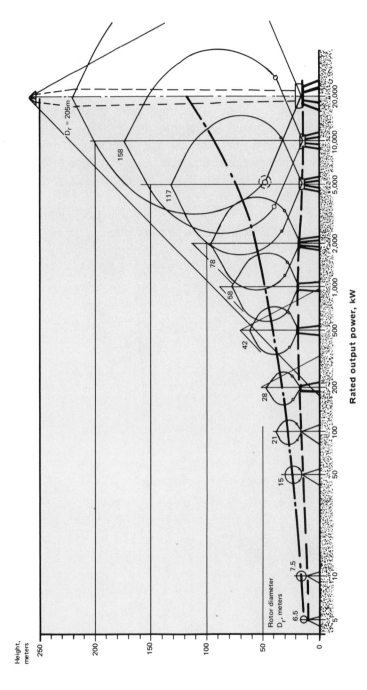

Figure 48. Typical family of vertical-axis, Darrieus-type wind turbines for average wind speeds of 17 mph.

In addition, the federal government is in the process of constructing and demonstrating a number of types of WEC systems for farm, home, and other dispersed applications, including the generation of electricity, the pumping of water for household use, stock-watering, and irrigation purposes, and the heating of working fluids to provide hot water and space heating and cooling. Other potentially important applications are the use of wind power for desalination of water, using reverse osmosis, and the production of both low temperature and high temperature process-heat for industry and agriculture.

Most of the types of WECS units that are being applied under this program are the two-bladed horizontal-axis designs, although multi-bladed horizontal-axis rotors and various types of vertical-axis wind machines are also being used.

Performance and testing standards are also under development for the various types of WECS units used in this program.

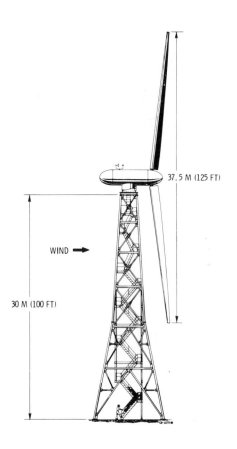

Figure 49. Dimensions of ERDA-NASA 100 kW test bed.

Figure 50. ERDA-NASA experimental wind-turbine generator — 100 kW test bed.

ERDA-NASA EXPERIMENTAL 100 kW UNIT

The first large-scale WECS unit that has been built under the federal program is located at the NASA-Lewis Research Center facility at Plum
Brook, Ohio, and was first put into operation in September 1975 (Figures
49 and 50).[14] It has a horizontal-axis, downwind rotor, 125 feet in diameter, designed for a rated capacity of 100 kW in 18 mph winds. The power
coefficient for the rotor blades, under these conditions, is 0.375, and the
power train efficiency is 0.75. Cut-in wind speed is 8 mph. The rotor is
designed to operate at a constant speed of 40 rpm in winds over 6 mph, by
changing the system load and the pitch of the blades. The blades are fully
feathered in wind speeds greater than 60 mph and the system is designed
to withstand wind velocities as high as 150 mph.

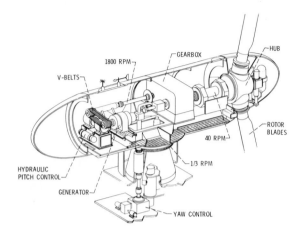

Figure 51. 100 kW wind-turbine drive-train assembly.

The rotor, transmission train, generator, and controls (Figure 51) are mounted on a bedplate on top of a 100 foot tower of pinned-truss design.

The original tower that was used for the Plum Brook machine was built with rather massive I-beams to provide stability, and had a stairwell leading to the top.

The controls for the Plum Brook machine consist of a pitch-change mechanism (Figures 52 and 53), as well as a yaw control mechanism at the top of the tower for rotating the bedplate at a speed of about 1/24 rpm. Torque is transmitted from the rotor hub to an electric alternator (Figure 54) through a gear box (Figure 55) with a speed ratio of 45/1. The alternator is rated at 125 kVA at 1,800 rpm and weighs about 1,425 lbs.

This initial 100 kW system was designed to serve as a test-bed for improved WECS components and subsystems and is being used to collect

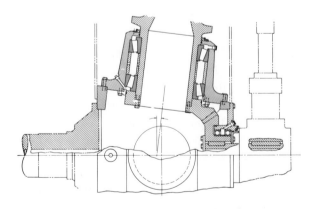

Figure 52. Hub- and pitch-change assembly.

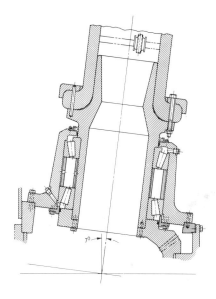

Figure 53. Blade root and bearing attachment.

125 KVA
480 V, Y-CONNECTED
60 Hz
1800 RPM
WEIGHT: 1425 LB
APPROX DIMENSIONS 2 FT DIAM, 3 FT LONG

Figure 54. Synchronous generator for ERDA-NASA 100 kW experimental unit.

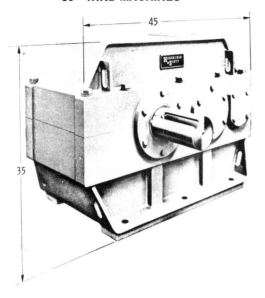 

RATIO   - 45:1
RATING   - 236 HP AT 1800 RPM
APPROX WT   - 4800 LB

Figure 55. Gearbox for ERDA-NASA 100 kW experimental unit.

performance data to be applied in the design of other wind generators of all sizes that will be built during the course of the WECS development and demonstration program. Performance data have been collected, including power output at various wind speeds; loads, stresses, and vibrations in components such as the blades, hub, and tower; and measurements of the stability and effectiveness of the control systems incorporated in the WECS units.

Another objective of the initial studies of the 100 kW system has been to identify WECS components and subsystems for which life-cycle costs and maintenance could be reduced, and to acquire a basis for estimating mass-production costs of future WECS systems of various sizes and types.

Early tests of the 100 kw Plum Brook unit indicated that the rotor blades, which in this case are downwind from the tower, were being highly stressed each time they rotated past the tower. It was found that transient stresses were being produced by an effect known as "tower shadowing," where the tower structure causes blockage and turbulence of the windstream. This results in a rapid decrease in the force imparted by the windstream as the blades rotate to a position in back of the tower, followed by a sharp increase in the force of the windstream on the blades as they emerge from the shadow of the tower.

To correct this problem the tower was redesigned by streamlining the support-beams and replacing the stairwell with a small elevator. The improved Plum Brook system was tied into the local electric utility grid in September 1977 and since then has been operating unattended, supplying power to the grid.

The lessons learned in the development of the Plum Brook system have been applied in the design of the second wind machine in this series. This unit, which was put into operation at Clayton, New Mexico in January 1978, has a streamlined tower with an elevator (Figure 56). It is rated at 200 kW, even though it has the same rotor diameter as the 100 kW Plum Brook machine, since it was installed at a site with a higher-velocity wind-regime than that at Plum Brook.

Machines of this type have also been installed at other sites including Block Island, which is off the coast of Rhode Island, and Culebra Island, which is east of Puerto Rico.

SYSTEM SIZES

Recent studies have indicated that there will probably not be any distinct economies-of-scale for wind-powered systems; i.e., the cost of energy generated by small-scale WECS is expected to be about the same to the consumer as the costs of energy generated by medium and large-scale WECS.[15]

In such cases, unit sizes are generally categorized in terms of output power as follows.

- Small-scale units: wind machines with rated power output capabilities up to 25 kW, in wind speeds of 20 to 25 mph.
- Medium-scale units: wind machines with rated power outputs between 25 and 250 kW, in wind speeds of 20 to 25 mph.
- Large-scale units: wind machines with rated power outputs above 250 kW, in wind speeds of 20 to 25 mph.

In addition to the medium-sized 100 and 200 kW WECS units of the type that have been installed at Plum Brook, Ohio and Clayton, New Mexico, the federal government is sponsoring the design, development, and demonstration of a number of other sizes of WECS, including a unit with a minimum rated output power of 1 kW for a wind speed of 20 mph (Table 2); a unit with a minimum rated output power of 8 kW for a wind speed of 20 mph (Table 3); and a unit with a minimum rated output power of 40 kW for a wind speed of 20 mph (Table 4). In addition, a number of small-scale commercial units for pumping water and generating electricity are now available (see the Appendix on suppliers).

In general, the capital cost of the towers represents a greater portion of the total system capital cost for the smaller systems than for the larger systems.[16] However, most of the smaller units are being used in dispersed

Figure 56. Wind machine with 125 foot diameter rotor installed at Clayton, New Mexico, in January 1978; 200 kW (23 mph rated wind speed).

Table 2. Design goals for 1kW WTG system

- Output — 1 kW minimum, 120v to 115v d.c. with regulation for charging 120v battery system
- Rated wind speed — 9 m/s (20 mph)
- Cut-in wind speed — minimize   with regard to power
- Cut-out wind speed — maximize   production and system cost
- Survival wind speed — 75 m/s (165 mph)
- System life — 25 years minimum
- Reliability — mean time between failures (MTBF) no less than 10 years
- Capital cost — $1500/kW maximum (based on 9 m/s output)
- Components — rotor, transmission, generator, control system
- Not included — tower, foundation, batteries

Table 3. Design goals for 8kW WTG system

- Output — 8 kW minimum, 60 Hz, 120/240v
- Installation — tie in with utility or gas generator independent of back-up power source
- Rated wind speed — 9 m/s (20 mph)
- Cut-in wind speed — minimize
- Cut-out wind speed — maximize
- Survival wind speed — 75 m/s (165 mph)
- System life — 25 years minimum
- Capital cost — $750/kW
- Components — rotor, transmission, generator, control mechanism, tower
- Not included — batteries, inverter, other such secondary components, foundations

Table 4. Design goals for 40kW WTG system

- Output
  Mechanical design
  - Horizontal turning shaft 440, 880, 1760 rpm ±1%
  Electrical design
  - 3 phase, 480 v ±5%, 60 Hz, tie in with utility grid
  - 3 phase, 480 v ±5%, 60 Hz, independent operation
  - 3 phase, 240 v ±5%, 60 Hz, independent operation
  - 3 phase, 480 v ±5%, 60 Hz, tie in of two or more WTG's as independent utility
- Rated wind speed — 9 m/s (20 mph)
- Cut-in wind speed — minimize
- Cut-out wind speed — 27 m/s (60 mph) minimum
- Survival — 56 m/s (125 mph) minimum
- System life — 30 years minimum
- Capital cost — $500/kW, based on output in 9 m/s wind
- Components — rotor, transmission, generator, control system, tower, power processing equipment
- Not included — foundations

Table 5. Major characteristics of G.E. WTG rated at 2,000 kW in 25 mph winds

Rated power (system)	2000 kW(e)
Rated wind speed	25 mph
Cut-in wind speed	11 mph at 30 feet
Cut-out wind speed	35 mph above ground
Survival wind speed	150 mph (at rotor)
Cone angle	12°
Inclination of axis	0°
Rotor speed	35 rpm
Blade diameter	~200 feet
Blade twist	11°
Airfoil	NACA 230XX
Blade-ground clearance	~40 feet
Life	30 years with maintenance
Environment	−31°F to +120°F

**Table 6. Mod-2 wind turbine baseline design features
and characteristics**

Rated power	2500 kw
Rated wind at hub	28 mph
Cut-in wind at hub	14 mph
Cut-out wind at hub	45 mph
Rotor diameter	300 feet
Rotor speed	17.5 rpm
Rotor blade materials	Steel/paper honeycomb
Generator speed	1800 rpm
Generator type	Synchronous
Gear box ratio	103
Gear box type	3-Stage planetary
Hub height	200 feet
Tower type	Steel shell
Capacity factor	0.41
Energy collected if $\overline{V}$ = 14 mph @ 30 feet	8,300,000 kWhr/year
Availability	0.92

applications, rather than in utility networks, which results in smaller costs for the interconnection of units and for the transmission and distribution of the electricity generated by the WECS. Moreover, the smaller units may prove to be more reliable and have longer lifetimes than the large-scale WECS. For the large-scale systems, rotor sizes will eventually be limited by the strength that can be achieved in the design of bearings, blades, and other stressed components.

The first of the large-scale WECS in the federal series is being developed by the General Electric Company (Figure 57). It will be a horizontal-axis wind machine having two blades with a swept diameter of about 200 feet. The rated output power of the system will be 2,000 kW for a wind speed of 25 mph. These and other design characteristics of the system are shown in Table 5. The first machine of this type is scheduled to be installed and operated at Boone, North Carolina by November 1978.

An even larger machine (Figure 58) is being developed by Boeing Engineering and Construction.[17] This machine will have a rotor with a swept diameter of at least 300 feet and will be designed for a site with an average wind speed of 14 mph. Major features and characteristics of the baseline design for this system are shown in Table 6. The rated output power of these machines will be 2,500 kW for a wind speed of 28 mph.

Figure 57. General Electric wind machine with 200 foot rotor, rated at 2000 kW for a wind speed of 25 mph. Located at Boone, North Carolina.

Figure 58. Boeing Engineering and Construction design for a wind machine with 300 foot rotor, rated at 2,500 kW in a wind speed of 28 mph.

The blades will be constructed of steel and paper honeycomb. The center part of each blade will have a fixed pitch and will be mounted on a teetering hub. The blade tips will have a variable pitch to provide speed control. Another unique feature of the baseline design is the tower, which will consist of a 187 foot shell with four tubular strut braces at the 50 foot level.

Kaman Aerospace Corporation is designing and fabricating a 150 foot rotor blade (Figure 59) under a NASA contract as part of the federal wind

Figure 59. 150 foot filament-wound, fiberglass and epoxy blade constructed by Kaman Aerospace.

energy program.[17] The main leading edge spar is filament wound glass fiber construction. The two afterbody panels are constructed of fiberglass and paper honeycomb. A steel adapter at the root-end of the blade has been used to attach the blade to the structural test fixture. Half the length of a football field and weighing approximately 12 tons, the composite fiberglass-epoxy blade is intended for use in megawatt-size wind turbines of the horizontal-axis type.

Potential advantages of composite wind-turbine blades over metal wind-turbine blades include low production costs, lighter weight, resistance to failure due to damage or fatigue, longer service lifetimes, ease of maintenance, and reduced radio/TV interference. After completion of an extensive structural test program on the blade in the fall of 1978, Kaman hopes to build a set of blades for dynamic testing on one or more of the large wind turbines being developed under the federal wind energy program.

New designs for large-scale wind machines are also being developed in other countries. One of the largest wind machines in the world today stands at a tower height of 53 meters (i.e., almost 174 feet) in Ulfborg, on the west coast of Jutland, Denmark (Figure 60).[18] The swept diameter of

Figure 60. Tvind wind machine, Ulfborg, Denmark; 2 MW (32 mph rated wind speed).

the three blades is 54 meters (about 177 feet). This machine has been designed and constructed by the three Tvind Schools of Denmark, consisting of a Seminary, a Commuter High School, and a Junior College. Together, these schools constitute a "university," with approximately 55 permanent instructors and about 400 boarding students. The instructors purchased the materials and equipment for this wind machine with their own funds, and the instructors and students have provided the labor required to build it. Construction was initiated in May 1975 and completed in January 1978. During that time, the schools built the foundation and tower, the nacelle (Figure 61), the blades (Figure 62), and the control monitoring equipment. The gear-box (Figure 63), which provides a mechanism for increasing the speed of the drive shaft between the rotoring hub and the electric generator, was unused when purchased by the schools. It was originally designed for use with a mine hoist. The electric generator is an eight pole, synchronous a.c. generator with a rated output of 2,200 kVA. It was purchased from a Swedish used-equipment dealer. The large yaw bearing that supports the nacelle, and the 6 foot diameter bearings at the base of each blade, which are used in changing the pitch of the blades, were originally designed for use in a large construction-crane, but were adapted for this purpose.

The concrete tower is hollow and houses a small elevator. The tower consists of circular cross-sections of three different diameters, varying

Figure 61A. Tvind wind-machine nacelle.

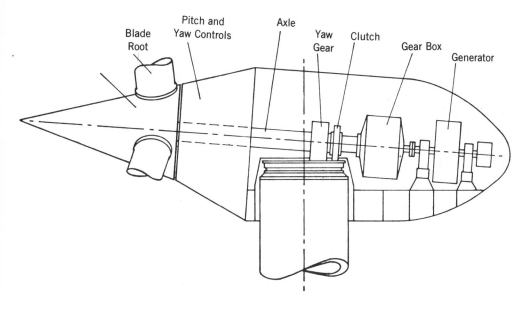

Figure 61B. Cut-away drawing of nacelle for Tvind wind machine.

Figure 62. Tvind wind-machine blade construction.

Figure 63. Tvind wind-machine gear box.

from 4.8 to 2.4 meters, with wall thicknesses of 0.3 to 0.5 meter. The rotor shaft is set at an elevation angle of 4° to help minimize the effects of tower shadow on the rotors. The rotors operate downwind from the tower and are canted at an angle of 9° from the vertical to assist in the yawing of the blades into the wind.

The tower was built with the aid of prefabricated forms that were intermittently raised by hydraulic jacks during construction, as the concrete was poured and hardened. This is similar to the manner in which dams, cooling towers, and other structures are built in the United States.

The nacelle (Figure 61) contains the main bearing for the rotor shaft, as well as the pitch control mechanism for the blades, the rotor shaft, the yaw bearing, the clutch, the gear box, and the generator.

The blades were fabricated from fiberglass ribbons, cemented with epoxy and laid in two half-forms—one for the frontside and one for the backside of the blade. The leading and trailing edges of the blades were built without forms. The inner structure of the blades consists of supporting spars and panels at about 1.5 meter intervals. The blade profile is that of the NACA 230 Series. These profiles have a very stable pressure center and a low drag coefficient. This type of blade profile continues to promote laminar flow even when its surface become roughened from corrosion,

Figure 64.   Conceptual drawings of off-shore installations.

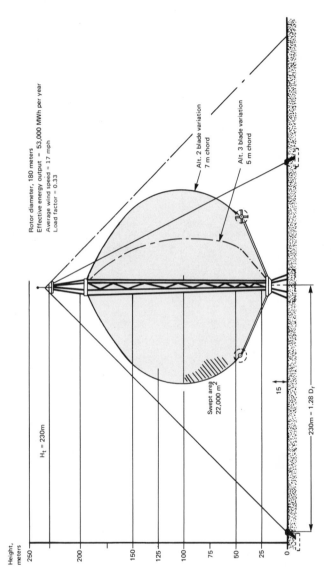

Figure 65. A design for a vertical-axis Darrieus wind turbine.

Figure 66. Model of icosahedron-supported Darrieus rotor on building.

Figure 67. Model of icosahedron-supported Darrieus rotor on water tower.

Figure 68. Model of contemplated tower with six Darrieus rotors.

wear, or contamination with adhering dirt. In order to assure that the natural vibration modes of the blades would be as high as possible, their contours were provided with a much greater twist than that for a design that would be aerodynamically optimum.

Taking all of these designs into account, it is probable that most future large-scale systems will be designed with rotors that are as light-weight as possible, to minimize the load on bearings, blades, tower, and other stressed components. The rotors will probably use lift forces rather than drag forces, because of the higher tip-to-wind speeds that can be achieved, resulting in higher power coefficients and smaller gear ratios.

Off-shore WECS (Figure 64) may be used in the future to take advantage of the higher-speed and more persistent winds available at such sites than are available over most land areas.[17]

Figure 69. Artist's conception of a tornado-type wind energy system.

Research is also being done at Sandia Laboratories, and in Canada and other places, to optimize the design and siting for a wide range of sizes of Darrieus-type rotors (Figures 65 to 68).[17]

Systems that incorporate towers designed to produce tornado-type vortices (Figure 69), using buildings or other types of natural baffles to increase the wind velocity and the pressure gradient across a wind turbine that extracts power from the system, are also under development.[17]

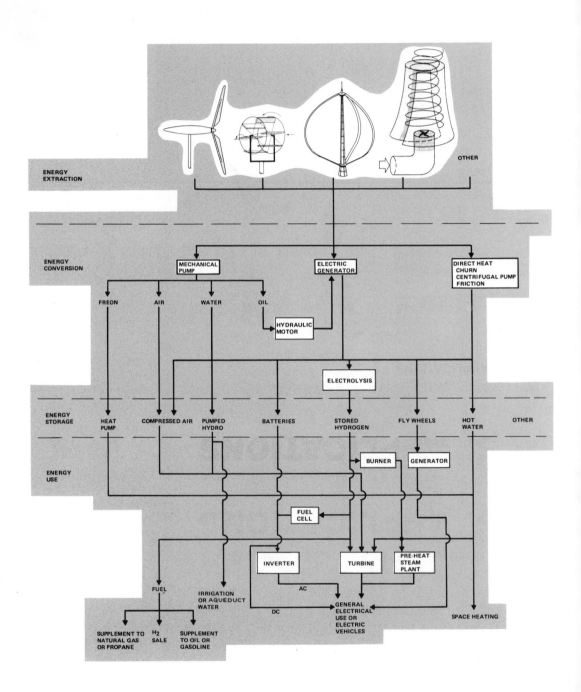

ENERGY EXTRACTION

OTHER

ENERGY CONVERSION

MECHANICAL PUMP

ELECTRIC GENERATOR

DIRECT HEAT CHURN CENTRIFUGAL PUMP FRICTION

FREON   AIR   WATER   OIL

HYDRAULIC MOTOR

ELECTROLYSIS

ENERGY STORAGE

HEAT PUMP   COMPRESSED AIR   PUMPED HYDRO   BATTERIES   STORED HYDROGEN   FLY WHEELS   HOT WATER   OTHER

ENERGY USE

BURNER   GENERATOR

FUEL CELL

INVERTER   TURBINE   PRE-HEAT STEAM PLANT

FUEL

AC

IRRIGATION OR AQUEDUCT WATER

DC

GENERAL ELECTRICAL USE OR ELECTRIC VEHICLES

SPACE HEATING

SUPPLEMENT TO NATURAL GAS OR PROPANE   H2 SALE   SUPPLEMENT TO OIL OR GASOLINE

5. APPLICATIONS

TYPES OF APPLICATIONS

Energy extracted from the wind is initially energy in the form of rotary, translational, or oscillatory mechanical motion. This mechanical motion can be used to pump fluids or can be converted to electricity, heat, or fuel. Some of the most effective applications are those that use energy derived directly from the wind, without further energy processing, conversion, or storage. However, if required, wind-derived energy can be converted to other forms of energy or can be stored through the use of compressed fluids, pumped-hydro systems, water-saver systems, batteries, hydrogen, flywheels, hot water, etc. Some energy is normally lost in each of these conversion or storage steps.

In any case, wind energy is one of the most flexible and tractable of all available energy sources, since the mechanical energy derived directly from the wind can be readily and efficiently converted to other forms of energy. The efficiency of converting wind-derived mechanical energy to heat or electrical energy is usually much higher, for instance, than the efficiency of converting solar or fuel-derived heat energy to mechanical or electrical energy since the efficiencies that can be attained when converting heat to mechanical or electrical energy are limited by the relatively low Carnot cycle efficiencies, which, even under optimum conditions, usually do not exceed 30 to 35%.

PUMPING APPLICATIONS

A typical wind-powered pumping application is one that might use a horizontal-axis wind machine. An example is the ancient jib-sail design that is still used to pump irrigation water in the Lasithi Valley on the island of Crete (Figure 70). There are so many windmills in this area that Lasithi is often called the "Valley of 10,000 Windmills."

Large numbers of water-pumping windmills have been used on American farms since the middle of the nineteenth century (Figure 22). A more modern version of an irrigation application is the wind-powered pumping system designed by the Brace Research Institute of Canada to pump water under high pressure for irrigation sprinklers (Figure 71). Such sprinkler distribution systems generally require a storage tank capable of providing water at 30 to 45 psi.

In other types of systems, gravity distribution from a small surface res-

Figure 70.   Crete, Valley of Lasithi with 10,000 windmills for pumping water.

ervoir is used. Wind-powered systems of this type have been developed by the U.S. Department of Agriculture. The Agricultural Research Service at Manhattan, Kansas has installed a 20 foot diameter, 30 foot high Darrieus rotor and wind-turbine, coupled to a shallow-well irrigation pump, for applications such as this; and the Agricultural Research Service at Bushland, Texas is operating a 37 foot diameter, 55 foot high rotor of the same type, in parallel with a 40 hp electric motor to drive a deep-well pump for irrigation applications.[17]

Other applications that are being developed include the pumping of water for aqueducts or for pumped-hydro storage of energy (Figure 72). In aqueduct systems, large-scale wind-driven units can provide power for the pumping of water from the main reservoir to auxiliary reservoirs in other parts of the aqueduct system. This can be done either by the direct mechanical pumping of the water or through the generation of electricity

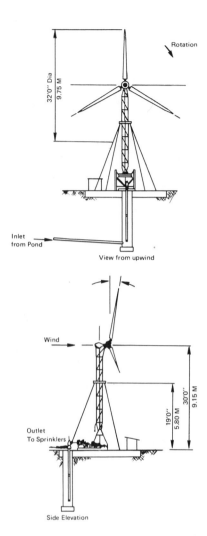

Figure 71. Prototype Brace Institute wind-powered pumping system.

by the wind units, and the subsequent use of this energy to operate electrical water pumps incorporated in the aqueduct system.[17]

In pumped-hydro applications, the wind units can be used to supply power to pump water from an auxiliary reservoir below a hydroelectric dam back into the main reservoir above the dam. This enables the water stored in the main reservoir to be replenished when the wind is blowing, thereby adding to the capacity of the hydroelectric system to generate base-load electrical power.

Wind power can also be used to compress air for use in various applications, including the operation of gas turbines for generating electricity during the peak-demand periods of a public utility system (Figures 73 and 74).[19,20] For this type of application, conventional gas turbines can be

Figure 72.   Wind generators used for pumped storage.

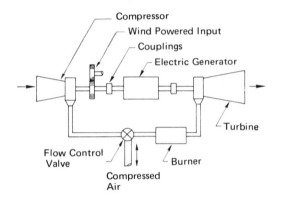

Figure 73.   Wind-assisted gas-turbine generating unit.

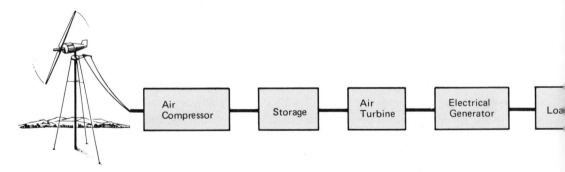

Figure 74.   System with compressed-air storage.

modified to separate the compressor, generator, and power stages by clutches. In one mode of operation, the motor-generator, operating as a motor and powered by a wind machine, drives the air compressor. The compressed air is fed into a storage tank or into a large cavern, aquifer, or depleted natural gas well. Under this mode, the power turbine is inoperative, and no fuel is consumed.

In a second mode of operation, when the demand for power exceeds the supply of the base-load utility system, the compressor is disengaged, and the power turbine is connected to the generator. The burner that drives the power turbine is fed fuel and compressed air from storage to generate power for the utility system.

The temperature of air is raised when it is compressed without loss of heat (i.e., adiabatic compression). If the air is stored in a well-insulated high-temperature container it will retain most of its heat (i.e., adiabatic storage). In this case, less heat will need to be added to the air, when it is eventually used to drive a turbine at a given efficiency, than if its heat has been allowed to escape from the storage container and the temperature of the air had been allowed to drop to the ambient temperature (i.e., isothermal storage). Adiabatic storage is obviously better, from the standpoint of energy conservation, than isothermal storage.

Wind-powered pumps can be used to desalinate water, using reverse osmosis units of the type that are being developed by Dupont and others. Recent improvement in the design of reverse osmosis membranes have resulted in units capable of desalinating seawater to potable water with only 500 parts per million of total dissolved solid, at operating pressures as low as 250 psi. This is well within the capabilities of wind-powered pumps.

Wind-powered pumps can also be used to save fuel and electricity by compressing the working fluids used in heat pumps for space-heating applications as discussed below.

DIRECT HEAT APPLICATIONS

Mechanical motion derived from wind power can be used to drive heat pumps or to produce heat from the friction of solid materials, or by the churning of water or other fluids, or in other cases, by the use of centrifugal or other types of pumps in combination with restrictive orifices that produce heat from friction and turbulence when the working fluid flows through them. This heat may then be stored in materials having a high heat-capacity, such as water, stones, eutectic salts, etc., or the heat may be used directly for such applications as heating and cooling of water and air-space for residential, commercial, industrial and agricultural buildings or for various types of industrial or agricultural process-heat applications.

An example of a wind-powered system designed for heating of a residential building is the "Wind Furnace" built by the University of Massachusetts (Figure 75).[21] This system uses a 32 foot diameter horizontal-axis wind turbine to generate electricity for a resistive heater that heats water in a storage tank. This water is then circulated through radiators in the house.

A somewhat similar system has been developed by the Agricultural Research Service at Ames, Iowa, that uses a Grumman 25 foot diameter, horizontal-axis wind turbine (Figure 76) to provide supplementary heat for a rural residence.[17]

The large wind turbine generator built by the Tvind Schools, as described previously, was designed primarily for heating the school buildings. This is done by using the electricity generated by the wind machine to provide resistive heating of water that is circulated through the buildings. This system uses a storage tank of 1,600 cubic meters capacity which can store the required thermal heat for eight to nine days, during periods in which calm winds might continuously prevail—including successive days with a mean temperature of −5°C. During the summertime, when space heating is not required, the electricity produced by the wind machine is sold to the local utility.[18]

Wesco, Ltd., of Surrey, England, has built several sizes of horizontal-axis machines ranging from 5 meters swept diameter to 18.3 meters swept

Figure 75. University of Massachusetts "Wind Furnace" home, Amherst, Massachusetts; 32-foot diameter rotor, rated at 29kW in 25 mph winds.

Figure 76. Grumman wind machine; 15 kW at 25 mph rated wind speed.

diameter which produce the equivalent of 10 kW to 165 kW at rated wind speeds. These machines use hydraulic transmission of power from the rotor to the energy converter and produce energy outputs consisting of heat or a combination of heat and electricity. A 165 kW wind machine of this type is being used to heat a one acre greenhouse in Hampshire, England.[22]

A home heating system that uses a wind-powered pump and a restrictive orifice to derive direct heat for a building, without first generating electricity, has been developed by the Browning Engineering Co. of Hanover, New Hampshire.[15]

The Dynergy Corporation of Laconia, New Hampshire is manufacturing a Darrieus vertical-axis wind turbine based on designs developed by the Sandia Laboratories. This has been supplemented by a "water twister," manufactured by All American Engineering Co. (Wilmington, Delaware), which heats water by mechanical agitation.

Examples of possible wind-powered agricultural process heat applications include greenhouse operations, crop drying, milk processing, food processing, refrigeration, frost protection, ventilation, and waste processing. In this regard, Virginia Polytechnic Institute has developed a wind turbine system to operate refrigeration equipment for an apple storage warehouse. Kaman Science Corporation has installed a 20 foot diameter, 30 foot high Darrieus rotor to drive refrigeration equipment for milk cooling at the Colorado State University experimental dairy farm. Cornell University has installed a vertical-axis wind turbine to provide heating of water for a dairy through the use of a hydraulic churn.[17]

Examples of typical industrial processes that might be able to use low temperature heat (i.e., up to approximately 350°F) produced by wind energy include the following.

- Production of inorganic chemicals, including borax, bromine, chlorine, caustic soda, potassium chloride, and sodium metal.
- Production of plastic materials and synthetics such as polyethylene, polyvinyl chloride, and polystyrene, for which approximately 45% of the process steam used is in the range of 212° to 350°F.
- Production of organic chemicals such as various types of alcohols and solvents, synthetic perfumes, flavoring materials, rubber processing chemicals, etc.
- Food processing such as meat preparation and packing, fruit and vegetable dehydration, wet corn milling, soybean oil milling, etc., for which approximately 80% of the process heat is used at temperatures less than 350°F. Of this, about 25% is used for process heating, 15% for cooking, and 62% for drying and dehydration.
- Textile processing, primarily steam or hot air, for drying, curing, and finishing of both yarns and textiles.

ELECTRIC GENERATION APPLICATIONS

Wind power can be used in centralized utility applications to drive synchronous a.c. electrical generators. In such applications, the energy is fed directly into power networks through voltage step-up transformers (Figures 77 to 81).

WECS units can be integrated with existing hydroelectric networks and used in a "water-saver" mode of operation. When the wind is blowing, electrical generation at the hydroelectric plants in the network can be reduced by an amount equal to that being produced by the WECS units. Thus, part of the network load that is ordinarily produced by the hydroelectric generators is supplied by the wind turbines. Under these conditions, some of the water that would have been used by the hydroelectric plant to supply the load is saved in the reservoir and made available for later use when the wind is not blowing. Additional hydroelectric generating facilities are provided at the hydro plant to allow the water that was saved in the reservoir to be used at a greater rate when the wind was not blowing, thereby providing a *firm* total generating capacity equal to the *firm* generating capacity of the hydro plant plus the *average* generating capacity of the wind-powered plant. The Bureau of Reclamation of the U.S. Department of the Interior has proposed such an installation in the vicinity of Medicine Bow, Wyoming, where, initially, 49 large-scale WECS units would be tied into the hydroelectric transmission network of the Colorado River Storage Project (Figure 82).[17] Other applications of

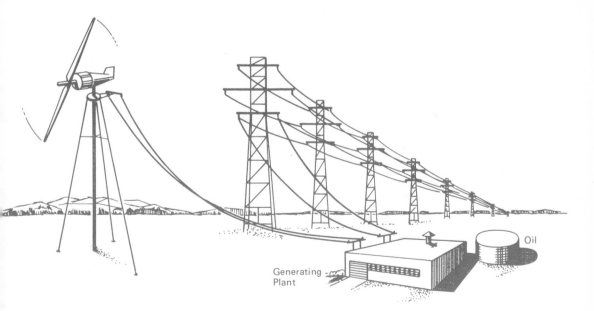

Figure 77. Wind generators with existing electric utility systems.

Figure 78. DOE/NASA, Clayton, New Mexico, wind machine in background, with
multi-bladed water-pumping wind machine in the right foreground.

this type are being considered by the Bonneville Power Administration
for use in the Pacific Northwest.

In dispersed applications, wind power can be used to generate d.c.
electrical power that, in turn, can be used for d.c. appliances or space
heaters, such as resistance heaters, or can be stored in batteries (Figure
83) and then inverted for use by a.c. loads.

There are a number of commercial units available for these types of
dispersed operations (Figures 84 to 93) and all can be used independently
or tied into a utility network using an induction generator or a Gemini
synchronous inverter (Figure 94). In this manner, about 150 dispersed
wind machines have been tied into a total of about 40 utilities in the last
few years.[23]

In centralized or dispersed applications requiring constantly available
sources of power, the energy can also be stored (Figure 95) in the form of
the mechanical motion of a flywheel (Figure 96) or as hydrogen and oxygen gases derived from the electrolytic dissociation of water (Figure 97).
The hydrogen and oxygen can be stored in liquid form in tanks, or in
gaseous form in tanks, caverns, aquifers, depleted natural gas wells, etc.
The stored hydrogen can be used either as a fuel for direct space heating

Figure 79. Wind Power Products Co. prototype of wind turbine generator with rotor 165 feet in diameter, designed to produce 3MW in a 40 mph wind.

or industrial process heat, or it can be reconverted to electricity through the use of fuel cells, gas turbine generators that burn hydrogen, or by other means.[19]

INTERCONNECTED SYSTEMS

There appear to be important advantages to using wind-derived energy in combination with energy derived from other sources, such as conventional fuels, sunlight, ocean thermal differences, bioconversion fuels, etc.

Figure 80. WTG wind machine on Cuttyhunk Island, Massachusetts; 100 kW (27 mph rated wind speed).

Figure 81. Energy Development Co., Hamburg, Pa.
45 kW (25 mph rated wind speed).

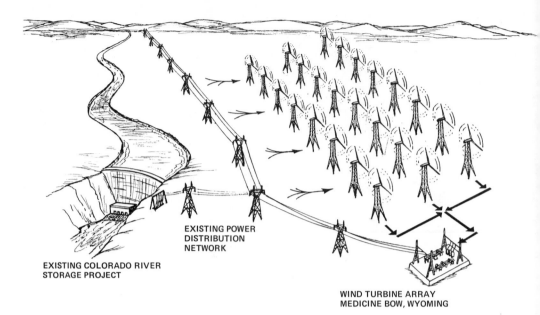

EXISTING POWER
DISTRIBUTION
NETWORK

EXISTING COLORADO RIVER
STORAGE PROJECT

WIND TURBINE ARRAY
MEDICINE BOW, WYOMING

Figure 82. Integration of wind and hydroelectric power.
Source: Bureau of Reclamation, DOI

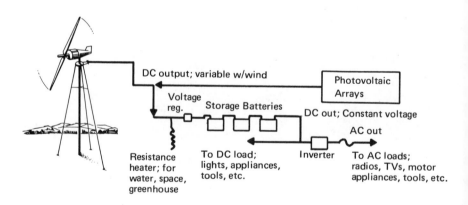

DC output; variable w/wind

Photovoltaic
Arrays

Voltage
reg.   Storage Batteries

DC out; Constant voltage

AC out

Resistance
heater; for
water, space,
greenhouse

To DC load;
lights, appliances,
tools, etc.

Inverter

To AC loads;
radios, TVs, motor
appliances, tools, etc.

Figure 83. System with battery storage.

Figure 84. Pinson wind-powered generator located on rooftop;
2 kW (20 mph rated wind speed).

Figure 85a. Enertech wind-powered generator with 13 foot rotor;
1.5kW (20 mph rated wind speed).

Figure 85b. Enertech machine undergoing tests on Mt. Washington, N.H., winter of 1979.

Figure 86. Restored Jacobs wind-powered generator;
2 kW (20 mph rated wind speed).

Figure 87. North Wind Eagle wind-powered generator;
2 kW (20 mph rated wind speed).

Figure 88. Electro wind-powered generator;
6 kW (26 mph rated wind speed).

Figure 89. Wind Engineering Corporation, wind-powered generator with 34 foot rotor; 23 kW (23 mph rated wind speed).

Figure 90. United States Windpower Associates, generator
with 10 meter rotor: 25 kW (26 mph rated wind speed).

Figure 91. United Technologies Research Center, windpowered
generator, 9.45 meter Composite Bearingless Rotor.
8 kW (20 mph rated wind speed)

Figure 92. Sencenbaugh wind-powered generator:
1 kW (22 mph rated wind speed).

Figure 93. Wind Power Systems, Inc., Storm Master—Model 10
with fiberglass rotor blades, 8 kW (18 mph rated wind speed).

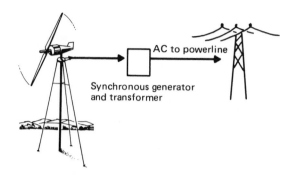

Figure 94. System with direct feed to main powerline.

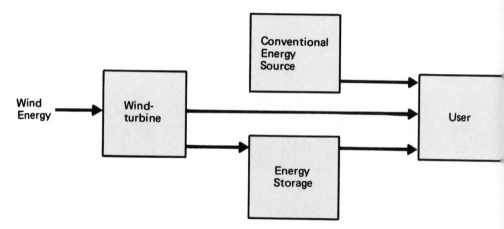

Figure 95. Basic wind energy conversion system with energy storage.

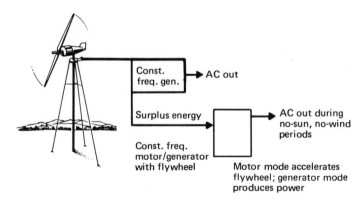

Figure 96. Alternative for converting and storing wind energy:
system with flywheel storage.

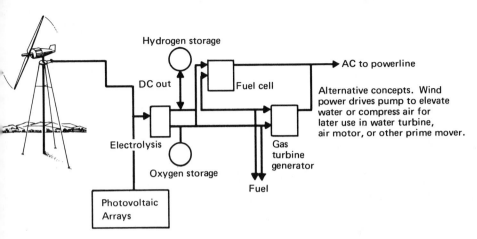

Figure 97. Alternative for converting and storing wind energy:
system with hydrogen storage.

Since the wind blows intermittently in most locations, there may be a need to store wind energy over long periods of time, perhaps up to ten days or more, if the energy is being used for isolated applications requiring continuous power. The cost of providing sufficient storage capacity for such applications can be reduced if the wind-derived power is interconnected with other sources of power. For instance, since in most locations the wind often blows when the sun is not shining, and vice-versa, a system using wind energy collectors and sun energy collectors (solar photovoltaic arrays or solar thermal collectors) in combination can be expected to require less energy storage capacity than systems that use these types of collectors separately.[20]

Large numbers of dispersed WECS units tied into the same grid network can also be used to reduce storage requirements for base load applications served by the grid, since wind speeds can vary considerably over large areas at any given moment in time.[24]

KEY:

Over 5000 kWh/kW

3750-5000 kWh/kW

2250-3750 kWh/kW

750-2250 kWh/kW

Under 750 kWh/kW

ANNUAL AVAILABILITY OF WIND ENERGY IN DIFFERENT PARTS OF THE WORLD

In Terms of Estimated Number of kWh/Year per Rated Kilowatt Output for
Wind Machines Designed for Rated Wind Speeds of 25 Miles per Hour

6. SITING

WIND DISTRIBUTION

Currently available maps showing patterns of average wind power over the United States provide only very rough estimates of the actual potential of this source at specific locations where wind machines might be sited (Figures 98A, and 98C).[25,26] For instance, many maps are based mainly on measurements obtained near ground level at airports. However, airport locations are, in general, purposely chosen to avoid sites where local topography might result in high wind speeds. Other pattern maps are produced by extrapolating high altitude wind measurements down to a standard height above ground level. An example of the latter type is shown in Figure 98B.[27] The shaded areas indicate where annual average wind speeds are estimated to equal or exceed 18 mph at 150 feet altitude over the United States. Many of these areas are near large population centers such as Buffalo, Cheyenne, Wichita, Amarillo, Boston, Chicago, San Francisco, Omaha, and New York. Others are areas served by large federal utility networks such as those of the Bureau of Relamation, The Bonneville Power Administration, and the Tennessee Valley Authority, or those of privately owned or municipal unilities, or REA cooperatives.

A rough interpretation of the average power available within the 18 mph contour surrounding the high Great Plains region indicates that, with conservative assumptions regarding the operating efficiencies and proper spacing of wind machines, the power that could be extracted from the winds in that region alone is expected to at least equal the present total United States demand for electrical energy, which in turn, is about 25 to 30% of the total United States demand for all forms of energy.[28]

Studies, at the University of Virginia, of winds in the coastal zones between 5 kilometers inland and 5 kilometers off-shore, indicate that maximum wind power is usually found in areas about 5 kilometers offshore and at altitudes about 50 meters above the sea surface.[17]

Only about 2% of the total solar energy intercepted by the earth is dissipated in the form of wind energy. However, the annual average power density of the near surface winds, as measured in a vertical plane perpendicular to the wind direction, is larger in many areas of the United States than the annual average solar power density, as measured in a horizontal plane (Figure 99).[15] Because of the global shape of the earth, a major part of the solar flux is received at the earth's surface within about 30° of the sun's latitude. Moreover, a significant part of the sun's energy is absorbed

105

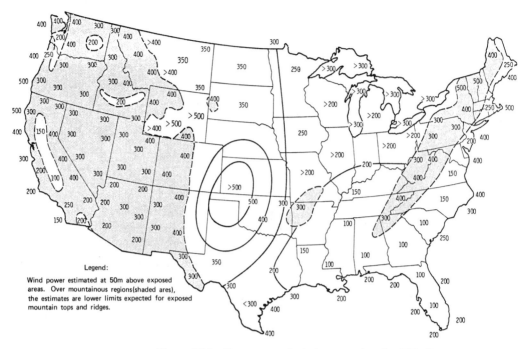

Legend:

Wind power estimated at 50m above exposed
areas. Over mountainous regions(shaded ares),
the estimates are lower limits expected for exposed
mountain tops and ridges.

Figure 98A. Mean annual wind power density (W/m²).

Source: Battelle Pacific Northwest Laboratories.

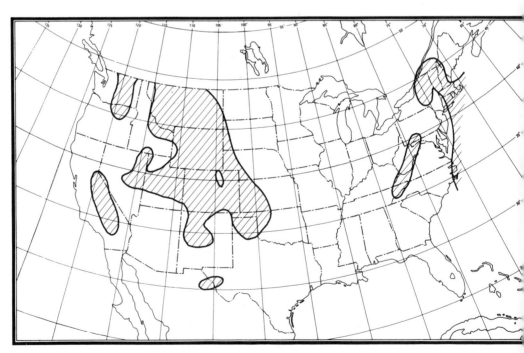

Figure 98B. Areas in the United States where annual average wind speeds exceed 18 mph at 150 feet elevation above ground level.

Source: General Electric Company

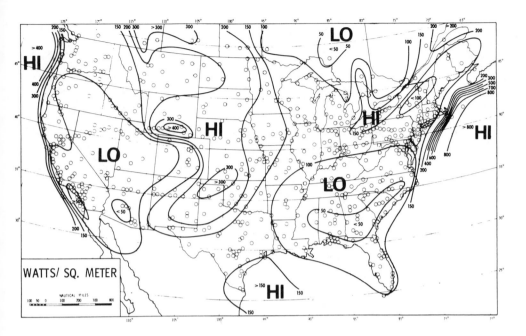

(After Reed — December, 1974)

Figure 98C. Available wind power density—annual average.

Source: Sandia Laboratories

in the atmosphere or is reflected back into outer space by the earth's cloud cover or by highly reflecting areas of the earth's surface, such as the oceans and seas. As a result, the wind and solar power densities also change with season (Figures 100 and 101).[15] This suggests that some of the most economic applications of wind power might be those that require large amounts of energy in the wintertime, such as for the heating of buildings. The advantage of using wind power for this particular application is magnified by the fact that heat losses in buildings are increased by the "wind chill" factor, and that wind is available for restoring heat to the building at the very time that wind chill is occurring—thus reducing the need for energy storage.

In some regions, shown by the heavily shaded portions of Figure 102,[15] it may often be economic to use a combination of wind and solar heating for buildings, depending on the orientation used for the solar collectors, the relative costs per unit area, and the efficiencies of the wind and solar collectors.

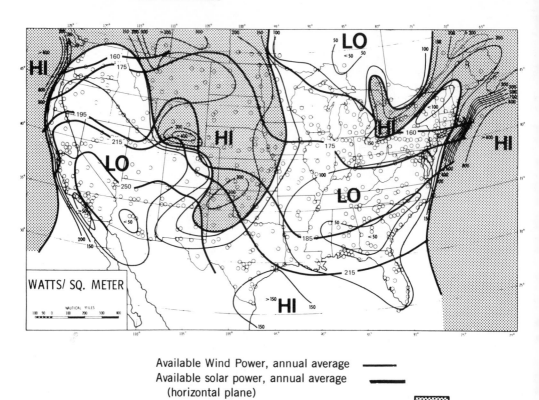

Available Wind Power, annual average ——————
Available solar power, annual average ——————
   (horizontal plane)
Areas in which wind power exceeds solar power ▨▨▨

Figure 99. Comparison of wind and solar power densities, annual average.

WIND CHARACTERISTICS

The amount of power available in a freely flowing windstream (Figure 103) of cross-sectional area, A, is equal to this area times the velocity of the windstream, V, times the kinetic energy of a unit volume of the windstream (or one-half the mass density of the air, ρ, times the subtended area of the windstream times the cube of the wind speed). The power per unit cross-sectional area of the windstream (i.e., the wind power density) is, therefore, equal to one-half the air mass density times the cube of the wind speed. As a result, the power densities of winds at sea level increase from about 5 watts per square foot for wind speeds of 10 mph, to over 140 watts per square foot for wind speeds of 30 mph, and to about 650 watts per square foot for winds of 50 mph (Figure 104). At higher altitude sites the air density is less, so the wind power density is lower for a given wind speed.

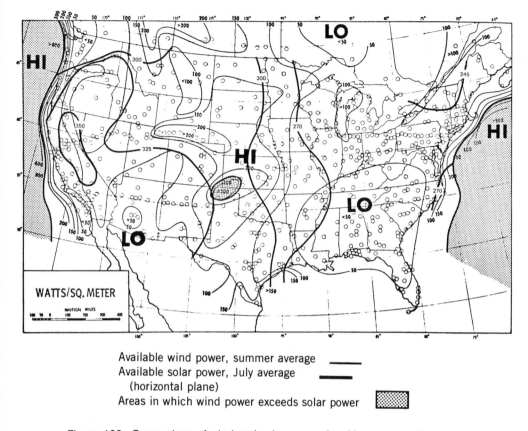

Available wind power, summer average   ────

Available solar power, July average   ━━━━
  (horizontal plane)

Areas in which wind power exceeds solar power

Figure 100. Comparison of wind and solar power densities, summertime.

The total available wind power in a freely flowing windstream will increase with the subtended area of the windstream (Figure 105). For instance, at sea level, a 20 mph windstream with a cross-sectional area of 100 square feet will contain about 4 kW of power, while over 40 MW of power will be available in a one-million square foot cross-sectional area of such a windstream.

The wind at a given site (Figure 106) usually varies frequently in direction (Figure 107) and its speed may change rapidly under gusting conditions (Figure 108).[26] Its average velocity also usually changes significantly with the season of the year. In many locations, the average wind velocities will be 30 to 35% higher in the winter months than in the summer months and the power density of the wind will be two to three times greater in the winter months than in the summer months. This is particularly true in areas with high annual average winds, such as in the northern latitudes on coastlines and in mountainous areas and the high Great Plains.

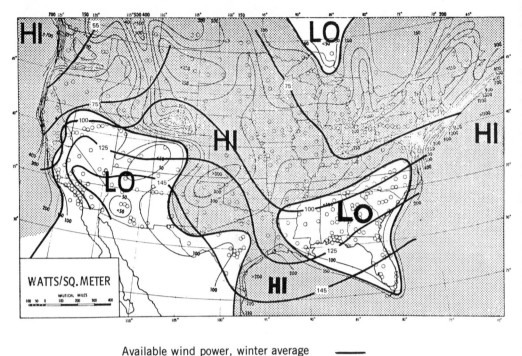

Available wind power, winter average   ——

Available solar power, January average   ━━
   (horizontal plane)

Areas in which wind power exceeds solar power

Figure 101. Comparison of wind and solar power densities,
wintertime.

Graphs showing the number of hours per year that the windstream attains a specified hourly mean wind speed at a given site are known as "annual average velocity duration curves" (Figure 109).

Curves showing the distribution of annual average wind power per unit subtended area, as a function of windspeed, are called "annual average power density distribution curves."

It is estimated by Sandia Laboratories that because of the cubic relationship between wind power and wind velocity, coupled with the fact that the wind gusts are seldom steady, the actual wind power available at a given site can be two or three times that calculated on the basis of average annual wind speeds at that site (Figure 110). Therefore, depending on the responsiveness of a wind machine to these changes in wind speeds, the estimated performance of the machine, if based on annual average wind speeds, may be quite conservative.

The "annual average wind *energy* density distribution" is equal to the annual average *power* density distribution times the number of hours per year that the corresponding wind speeds occur.

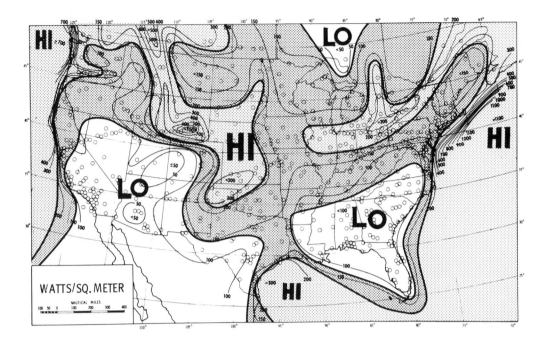

Available wind power, winter average
Available solar power, January average
(horizontal plane)
Areas in which wind power exceeds solar power

Figure 102. Areas favorable for both wind and solar heating applications,
wintertime.

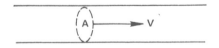

POWER = (VOLUMETRIC FLOW RATE)
x(KINETIC ENERGY PER UNIT VOLUME)

$$P = (AV) \times \left(\frac{\rho V^2}{2}\right)$$

$$\text{or } P = \frac{\rho A V^3}{2}$$

Figure 103. Power in windstream.

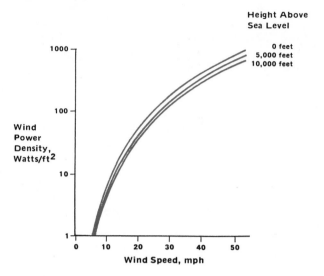

Figure 104. Change of windpower density with wind speed and altitude.

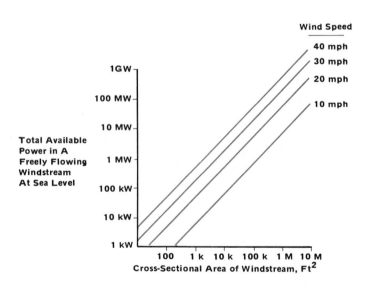

Figure 105. Power in windstream vs. cross-sectional area and wind speed.

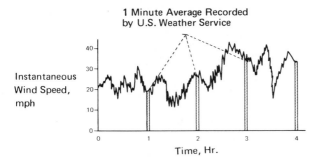

Figure 106. Typical wind speed record.

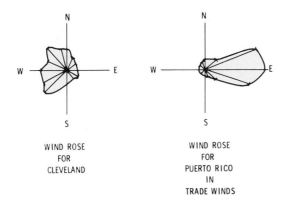

Figure 107. A wind rose shows the hours per year that the wind blows from each direction.

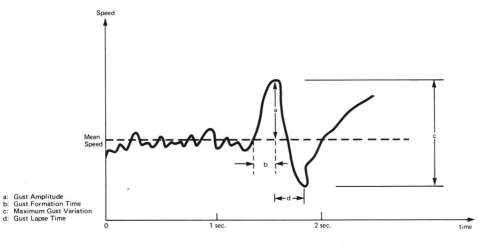

Figure 108. An example of gusting.

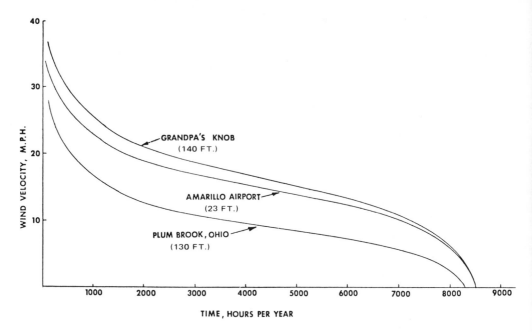

Figure 109. Annual average velocity duration curves for three sites.

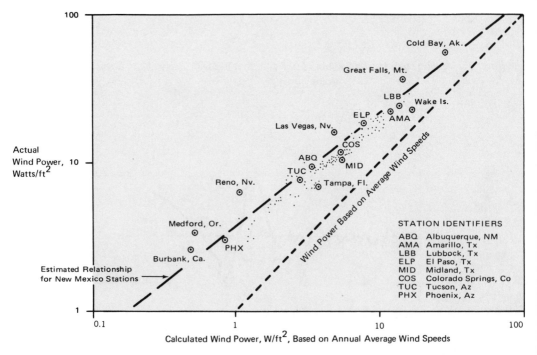

Figure 110. Comparison of actual and calculated wind power at different sites.

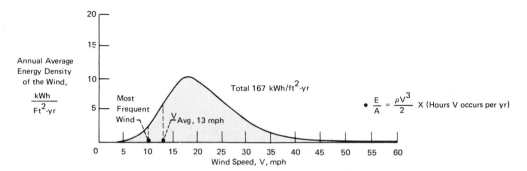

Figure 111. Typical distribution of annual average energy density of winds of various speeds.

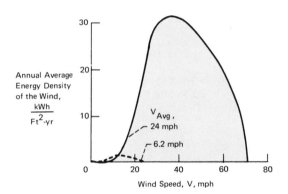

Figure 112. Typical distribution of annual average energy density of winds for sites with low and high annual average wind speeds.

Plots of the distribution of the annual average energy density of winds of various speeds at a given site show that most of the energy content of the wind is at speeds above the average wind speed (Figures 111 and 112), but that the contribution to the total annual average content of winds of all speeds is usually small for winds of speeds greater than about three times the average wind speed.

The geographical distribution of average wind speeds is quite skewed (Figures 113 and 114), with some protected sites in the interior land areas showing almost no wind, while other areas, particularly in northern coastal and mountainous areas and the high Great Plains, record average winds up to 20 to 25 mph at 20 meters above ground level, and 28 to 35 mph at 40 meters height, based on 758 sites in the 50 states as well as

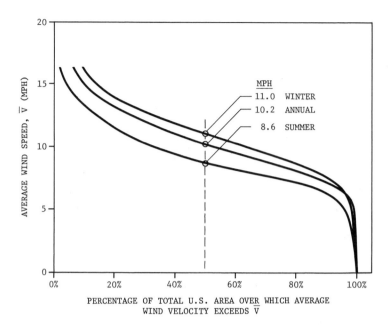

Figure 113. Estimated geographical distribution of average wind speed at 20 meters above ground level.

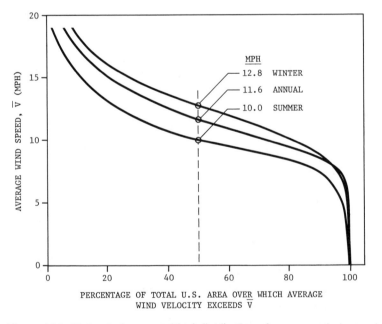

Figure 114. Estimated geographical distribution of average wind speed at 40 meters above ground level.

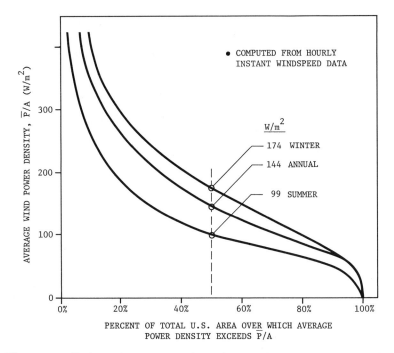

Figure 115. Estimated geographical distribution of average power density
at 20 meters above ground level.

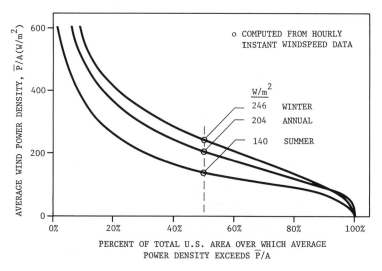

Figure 116. Estimated geographical distribution of average power density
at 40 meters above ground level.

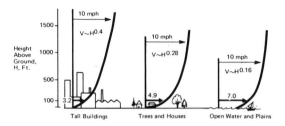

Figure 117. Effect of ground roughness on vertical distribution of wind speeds.

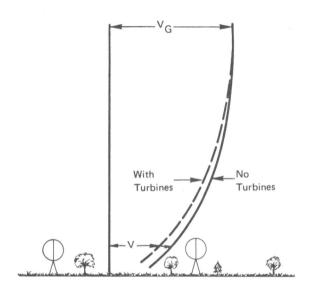

Figure 118. Effect of turbine array on wind velocities at various heights.

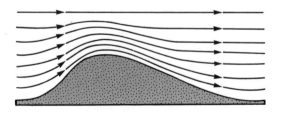

Figure 119. Acceleration of wind over hill.

WIND FEATURES	EFFECTS
• Unsteadiness	• Wind energy is unsteady and sometimes zero; energy storage or backup system may be required for applications requiring a constant source of power.
• Omni-directionality	• To maximize the energy output, horizontal-axis type rotors must always be vectored into the wind.
• Spatial non-uniformity due to ground roughness and terrain variations	• Wind energy units must be ruggedly built to resist gusts and high winds.
• Horizontal variations from place to place	• Optimum sites may be hard to find in contoured areas.
• Vertical variations	• Even for level terrain, rough ground surface usually results in turbulence and low winds under 30 feet; taller towers are required on rough ground than on smooth ground to reach the levels at which laminar flow exists.

some sites in the southern parts of Canada, where wind measurements have been made by the Weather Bureaus of the two countries over a period of decades. Most of these measurements were originally made at altitudes of 7 to 10 meters and have been extrapolated to 20 and 40 meters, assuming that wind speed at these locations varies with height to the one-sixth power.

The geographic distribution of average wind power density is even more skewed (Figures 115 and 116) because power density varies as the cube of the wind speed.

CHOICE OF SITE

Care should be exercised in choosing a suitable site for a wind machine since there is a significant amount of shear and compression in a normally horizontal windstream as it passes over the topographical contours and the rough surface of the earth. This shear results in lower wind speeds near the surface than at heights great enough for free wind flow to occur. In particular, the free flow velocity, at heights large enough to be unaffected by the surface shear, is usually significantly larger than that of winds either at the surface or at standard anemometer heights of 20 to 30 feet, where the wind is normally measured. A common rule of thumb is that wind speeds near the earth's surface will increase as the one-seventh power of the height above the earth's surface over open water and flat plains (Figure 117).

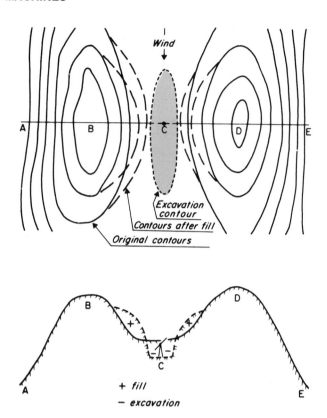

Figure 120. Proposed type of terrain modification for the purpose of augmenting average wind speeds.

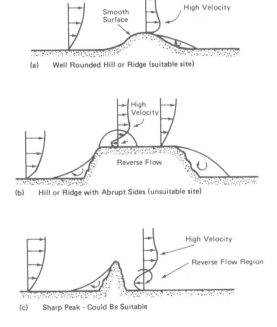

Figure 121. Care should be exercised in selecting suitable site.

SUMMARY OF FEATURES OF SUITABLE SITE

- High annual average wind speed (consult local National Weather Service Station)

- No tall obstructions upwind for a distance depending on the height

- Top of smooth well-rounded hill (with gentle slopes) on flat plain or island in a lake or sea

- Open plain, open shoreline

- Mountain gap that produces a funneling

Figure 122. Summary of features of suitable site.

However, the wind shear, and consequently the available wind power at a given altitude, is also affected by the roughness of the earth's surface in a given location. If the area contains buildings, trees, wind machines, or other obstacles, the variation of wind speed with altitude above ground level is usually greater for these obstructed areas than for the case of open water and flat plains (Figures 117 and 118).

Some wind tunnel tests have been made in Sweden[29] and in other places to determine the minimum permissible spacing of wind rotors in a given geographical area to prevent significant interaction of the rotors that could result from the windstream turbulence that is normally created by the rotors. From this work, it appears that the spacing should be at least 6 rotor diameters, and that an 8 to 12 diameter spacing may be most desirable.

Another significant effect is that the streamlines of a windstream are compressed and its flow accelerated as it passes over a hill or through a narrow valley (Figures 119, 120, 121, and 122). It is often possible to increase the average power output of a wind machine by siting it in such a way as to take advantage of the increased average wind speed that results from anomalies such as these.

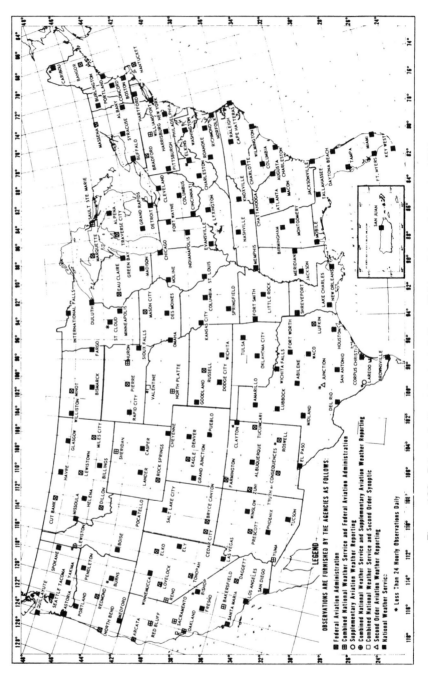

Figure 123.　Principal climatological stations (24-hourly) as of January 1974.

Figure 124. Instrument tower for collecting wind data.

WIND SURVEYS

The National Climatic Center, NOAA, collects wind characteristics data from over 700 weather stations throughout the United States from which average wind speeds and directions and monthly and annual average wind energy and wind power distributions have been obtained (Figure 123).[26] However, because of local anomalies, it is usually worthwhile to make detailed wind surveys before choosing a site for a wind machine. An appendix includes the names of private firms that provide such services.

Typical wind measurements at potential sites for wind machines usually require the following.

- Instrumentation
  3-cup anemometer and wind direction sensor
  Height of instruments (Figure 124): 30 feet for preliminary data; 50 to 150 feet for longtime data
- Data recording systems
  Strip chart
  Magnetic tape
- Type of data
  Wind speed and direction—hourly averages
- Data reporting
  Wind frequency curves
  Daily, weekly, monthly

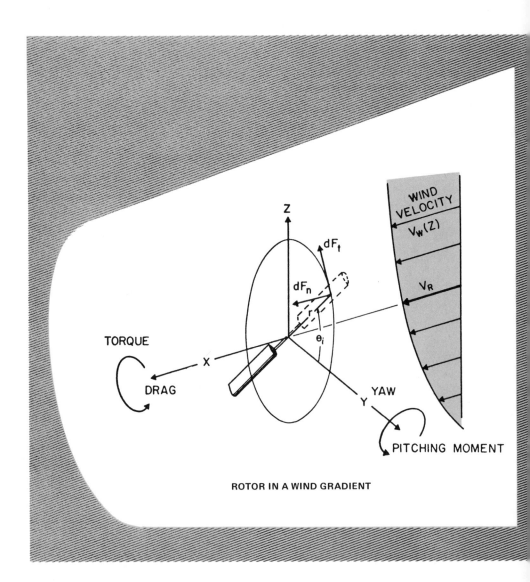

ROTOR IN A WIND GRADIENT

7. PERFORMANCE

POWER EXTRACTION

The power, P, that can be extracted from a windstream by an unshrouded wind turbine (Figure 125) is equal to the efficiency, η, of the system used to extract the power, times the wind's volumetric flow rate at the turbine $V_T A_T$, times the sum of the change in pressure energy, Δp, plus the change in kinetic energy, ΔQ, of a unit volume of air that passes through the turbine:

$$P = V_T A_T [\Delta p + \Delta Q].$$

For a conventional wind turbine operating under optimum steady state conditions in an initially free flowing windstream of constant initial velocity, V_0, the velocity of a unit volume of air will decrease monotonically as it approaches the turbine (Figure 126), passes through it, and starts to recede from it. As the air recedes farther from the turbine, it will receive kinetic energy from the surrounding winds and its velocity will increase until it again reaches its initial velocity, V_0.

The sum of the change in pressure energy and the change in kinetic energy of the unit volume of air will be constant as it approaches the turbine upstream. As it nears the turbine, the pressure energy will increase,

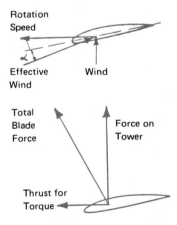

Figure 125. Essential wind turbine blade relationships.

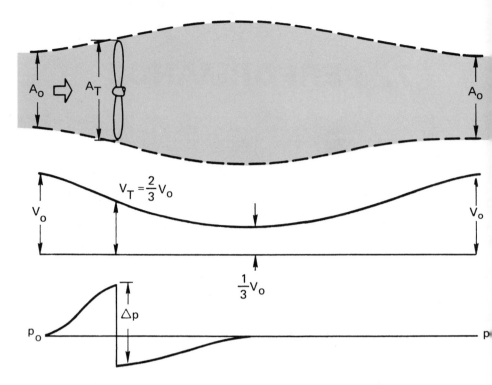

Figure 126. Optimal performance of conventional wind turbine.

and the kinetic energy will decrease, until the pressure reaches a maximum at the interface with the turbine. As the air passes through the turbine, it will impart kinetic energy to the turbine, and the pressure energy will drop to a level below atmospheric pressure. As it recedes from the turbine, its pressure will rise until it again reaches atmospheric pressure. After passing the turbine, its kinetic energy will continue to decrease until its pressure again reaches atmospheric pressure. After this, its kinetic energy will increase, as it receives kinetic energy from the surrounding winds, until its kinetic energy is the same as that of the surrounding wind.

The cross-sectional area of the windstream that passes through the system will be inversely proportional to its velocity. Its streamlines will expand as it approaches the turbine, passes through it, and starts to recede from it. As the disturbed flow receives kinetic energy from the surrounding winds, the effects of the turbine will diminish through the dispersal of the disturbance in the windstream.

The power coefficient, C_p, of such a system is defined as the power delivered by the system divided by the total power available in the cross-sectional area of the windstream subtended by the wind turbine.

There are significant trade-offs to be considered in the design of wind turbines (Figures 127 and 128). Turbines with a small total blade area will

$$\text{Power Coefficient, } C_P = \frac{\text{Power Delivered}}{\tfrac{1}{2}\rho\,A_T\,V_o{}^3}$$

Turbine Blade Area	Flow Rate	Pressure Drop at Turbine	Power Delivered	C_P
Small	High	Small	Low	<0.59
Large	Low	Large	Low	<0.59
Optimum	Optimum	Optimum	Optimum	0.59

Figure 127. Ideal power coefficient for various turbine blade areas.

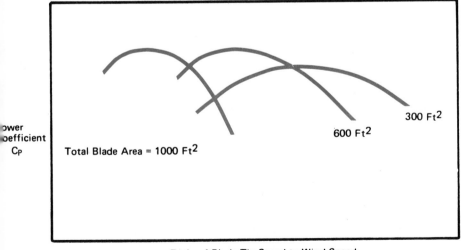

Power Coefficient C_P

Total Blade Area = 1000 Ft^2

600 Ft^2

300 Ft^2

Ratio of Blade Tip Speed-to-Wind Speed

Figure 128. Power coefficient vs. rotational speed and blade area.

maintain a high volumetric flow rate of the airstream, but the pressure drop will be relatively small, and the resulting power output and power coefficient will be small.

For turbines with a large total blade area, the change in pressure and the reduction in windstream velocity will be large, but the volumetric flow rate of the windstream will be low. This will also result in a low power output and a low power coefficient.

The optimum total blade design, therefore, is one that maximizes the product of the volumetric flow rate and the pressure drop across the turbine.

It can be shown from momentum theory that the maximum amount of energy that can be extracted by a wind turbine from a windstream, in the process described above, is eight-ninths of the kinetic energy of the windstream passing through it. Under these conditions, the maximum loss in wind speed will be two-thirds of the initial velocity of the windstream, V_O.

For a system of 100% efficiency, the maximum power density that can be extracted by the turbine from the windstream in the above process will be:

$$\frac{P_{max}}{A_T} = \frac{2}{3} V_0 \left[\frac{8}{9} \frac{(\rho V_o^2)}{2} \right] = 0.593 \frac{\rho V_o^3}{2}$$

where

$\dfrac{\rho V_o^3}{2}$ = the ambient power density of a unit volume of the windstream.

The factor 0.593 is known as the Betz coefficient (from the name of the man who first derived it). It is the maximum fraction of the power in a windstream that can be extracted by a turbine in the windstream.

TYPICAL POWER COEFFICIENTS

The power coefficient of an ideal wind machine rotor varies with the ratio of blade tip speed to free-flow windstream speed and approaches the maximum value of 0.59 when this ratio reaches a value of 5 or 6 (Figure 129).[30]

Experimental evidence indicates that two-bladed rotors of good aerodynamic design, running at high rotational speeds (i.e., where the ratio of the blade tip speed to free-flow speed of the windstream is 5 or 6) will have power coefficients as high as 0.47 (Figure 130).[30]

Likewise, a Darrieus rotor can be expected to reach a maximum power coefficient of about 0.35 at a ratio of peripheral speed to windstream speed of about 6 (Figure 129). Other designs have been found to have

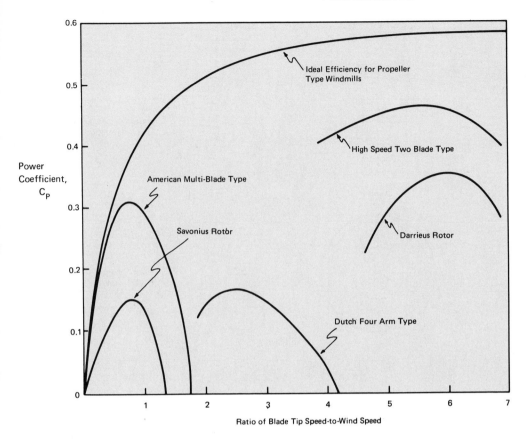

Figure 129. Typical performances of wind machines.

lower maximum power coefficients which occur at lower speed ratios than for the high speed two-bladed horizontal-axis rotors and the Darrieus vertical-axis rotors (Figures 129 and 130).[30]

OUTPUT PERFORMANCE

The torque developed by a horizontal-axis rotor blade of fixed pitch angle varies with both the shaft speed and the wind velocity (Figure 131). If the rotational speed of a blade is too slow in a windstream of a given velocity, the blade will stall, and the output torque of the wind machine will decrease. Therefore, to develop maximum power output from a windstream, as the speed of the windstream varies, either the pitch-angle of the blade or the rotational speed of the blade must be varied. Many of the more modern wind machines are designed with variable-pitch blades. For these, a control mechanism can be provided that will adjust the pitch

ROTOR TYPE	c_p MAX	COMMENTS
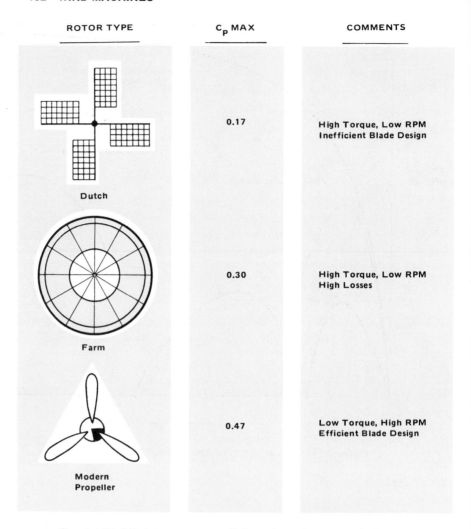 Dutch	0.17	High Torque, Low RPM Inefficient Blade Design
Farm	0.30	High Torque, Low RPM High Losses
Modern Propeller	0.47	Low Torque, High RPM Efficient Blade Design

Figure 130. Maximum power coefficients for various rotor designs.

of the blades to maintain a constant rotational speed as the wind speed changes or the output load varies.

At wind speeds lower than the rated value, the rotor speed must vary with the wind speed in order to extract maximum possible power. This does not match the optimum operating conditions for either synchronous or induction a.c. generators, if these types of generators are being driven by the wind machine rotor. This mismatch problem has led to several interesting concepts for interfacing the rotor with its electrical output. One approach is to allow the rotor speed to vary optimally with the wind speed and to employ a variable-speed, constant-frequency generating system to obtain the constant-frequency power needed for an electric utility grid. Both differential (e.g., mechanical or electrical frequency changing de-

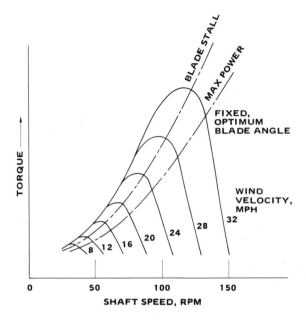

Figure 131. Speed-torque curve for typical optimized wind rotor.

vices) or non-differential (e.g., static frequency changers or rotary devices) may be used to accomplish this.

In other cases, such as the initial ERDA-NASA 100 kW wind machines installed at Plum Brook, power output at wind speeds greater than its rated value of 18 mph has been sacrificed to simplify system design (Figure 132). For a power coefficient, C_p, of 0.35, this machine has a theoretical maximum power output as shown by the dotted line in the diagram to the left and would reach an output of 4935 kW at wind speeds of 60 mph. However, a synchronous a.c. generator with a fixed-ratio gear train has been chosen for this design. This type of generator must be operated at constant input speed (i.e., fixed wind rotor speed) to maintain synchronization with the current in the utility grid to which it is inter-tied. Aside from friction losses, the theoretical maximum power output is main-

wind speed of 18 mph, by varying the output load on the system to maintain a constant rotor speed of 40 rpm. For wind speeds above 18 mph the rotor speed is maintained at 40 rpm by varying the pitch of the rotor blades. In this manner, a constant power output of 100 kW is maintained between the rated wind speed of 18 mph and the cut-off wind speed of 60 mph, at which point the rotor blades are feathered to protect the machine in high winds. This type of design does not result in large losses of available wind power, since, for example, when the average wind speeds are about 12 mph, the annual energy content of the wind is low for wind speeds below 8 mph and above 30 or 40 mph, as seen in Figure 132.

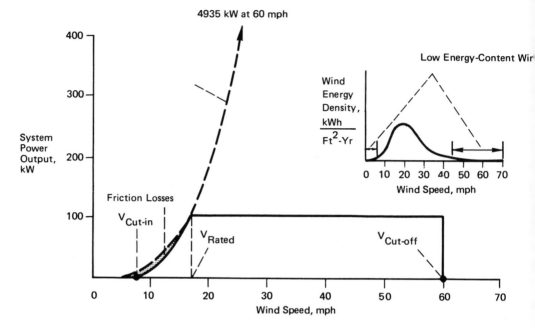

Figure 132. Power output of 125 foot diameter rotor, rated at 100 kW in winds of over 18 mph.

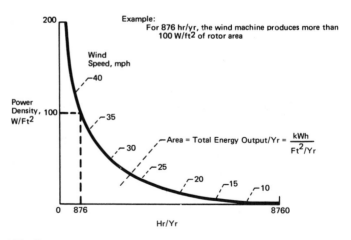

Figure 133. Theoretical power density duration curve for a WECS with $C_p = 0.35$.

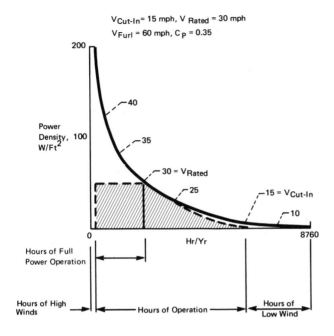

Figure 134. Actual annual power density output of a WECS.

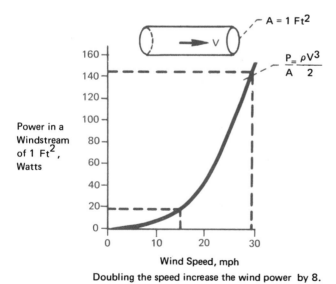

Doubling the speed increase the wind power by 8.

Figure 135. Power density of a windstream for various speeds.

A power density duration curve shows the number of hours per year that the available winds at a given site will provide a power density of a specified amount for a wind machine of a specified power coefficient located at that site. The area under such a power density curve represents the total amount of available energy per year, per unit of swept area of a wind machine located at that site. For example, theoretically, a WECS with a power coefficient of 0.35, located at a site with the power density duration curve shown in Figure 133, would produce more than 100 watts of power output for each square foot of rotor area during 876 hours of the year.

The actual power density of such a WECS with a synchronous a.c. generator driven by a gear train with a fixed ratio would be less than for the theoretical case shown. For instance, if such a WECS was designed for a cut-in wind speed of 15 mph, a rated wind speed of 30 mph, and a cut-off wind speed of 60 mph, its actual output power density would vary according to the dotted line in Figure 134, and its actual energy density output per year would be represented by the shaded area under this dotted line.

Since the power density of a windstream varies with both the velocity of the windstream and the swept area of the rotor, the power output of a wind machine can be increased both by locating it in areas of higher wind velocity (Figure 135) and by increasing the diameter of its rotor (Figure 136).

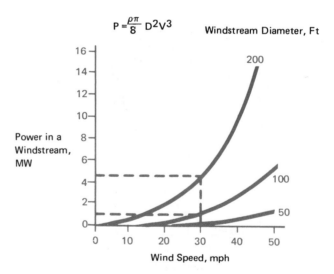

Figure 136. High power levels are available using large rotor diameters.

However, the specific size and the siting of such a wind machine should be chosen to minimize the cost to the consumer of the energy produced, and must take into account such factors as the transmission and distribution costs of the system and the physical limitations on rotor blade size.

EXTRANEOUS LOADING

Various types of extraneous loadings that occur in the operation of any wind machine can have a significant effect on blade motions (Figure 137). These types of loadings include gyro forces, blade balance, wind shear, wind shifts, wind gusts, gravity forces, and tower shadow (Figure 138). These loadings can cause cyclic motions and vibration in the blade, the tower, the bearings, and other system components and can have a serious impact on system reliability, lifetime, and performance. Such loadings, therefore, need to be carefully analyzed when designing wind machines. They are liable to be more serious for large wind machines than for small ones.

DUCTED SYSTEMS

Various types of ducted systems using wind concentrators, diffusers, and other types of man-made or natural structures can be designed to increase the amount of wind energy received by a system, thereby increasing both the wind speed and the pressure drop across the wind turbine (Figure

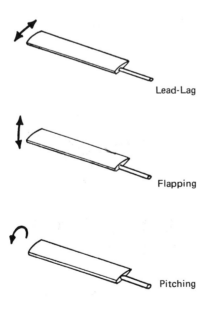

Lead-Lag

Flapping

Pitching

Figure 137. Rotor blade motions.

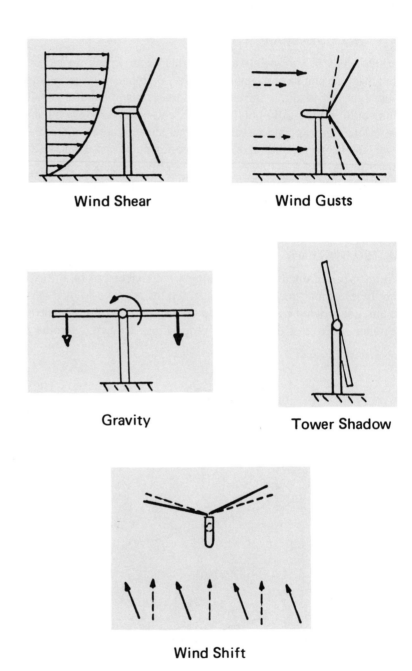

Figure 138. Causes of extraneous system loads.

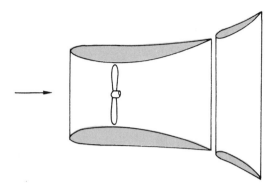

Figure 139. Cross-section of ducted turbine with diffuser.

139). This, in turn, increases the amount of kinetic energy that can be extracted from the wind by a turbine of a given diameter. Such measures can be used to decrease the size of a wind turbine required for a given power output and increase the rotational speed of the turbine. This decreases the amount of gearing required to interconnect the turbine and an electrical generator or other high-speed output device.

Ducted systems using wind concentrators and diffusers have been built that increase the wind speed through the turbine by a factor of 1.5 compared to the free-flow wind speed and produce up to 3.5 times the power generated by an ideal wind turbine having a diameter equal to that of the ducted turbine and operating at the same ambient free-flow wind speed. In practice, specific designs have to be analyzed to determine if the additional power output outweighs the cost of the diffuser. In general, inlet ducts or entry cones have not been found to be very effective, whereas exit diffusers are still the subject of considerable research.

VORTEX GENERATORS

Both unconfined and confined vortex generators that spin the wind to increase the power output of a turbine located in or near the vortex are being studied. Unconfined vortex systems are under examination at both the Polytechnical Institute of New York and West Virginia University.[17] Such systems use wing-like structures to deflect the wind and to create an unconfined vortex around the turbine. It is estimated that such systems can be designed to provide up to six times the power output of a conventional wind machine with the same rotor diameter, located in a freely flowing windstream with the same wind speed as that in which the unconfined vortex system is located.

A confined vortex system is being developed by Grumman Aerospace Corporation, in which the presure drop across a ducted turbine, and the wind speed through it, are further enhanced by using additional ambient wind to generate a confined tornado-like vortex in a tower located at the exit of the duct (Figure 140).[17]

Such confined vortices can be created by a variety of structures designed to collect the wind over a wide area and spin it as it enters an enclosure, such as a tower, with a helical or circular cross-section, interconnected to the exit of the duct in which the turbine is located. For a typical system, the height of the tower might be three times the diameter of the tower or nine times the diameter of the turbine (Figure 141).

For such a tower, the speed, V_0, and pressure, p_0, of the wind entering the tower may be larger than those of the ambient wind, depending on how effectively the ambient wind is ducted into the tower. Inside the tower, the pressure and speed of the spinning wind near the tower wall will be approximately those of the entering wind. Dr. James Yen, of the Grumman Aerospace Corporation, predicts, on the basis of aerodynamic theory, that the angular velocity of the wind in a large-scale vortex can reach a maximum value of about ten times that of the angular velocity of the wind at the tower wall, at a radius that is about one-seventh the inside radius of the tower wall (Figure 142).[12] He indicates that the vertical velocity of the wind inside the tower will be close to zero at the tower wall

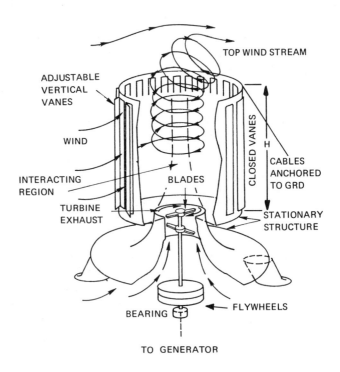

Figure 140. Grumman ducted turbine with vortex generator.

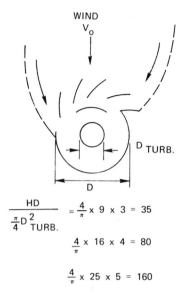

$$\frac{HD}{\frac{\pi}{4}D^2_{TURB.}} = \frac{4}{\pi} \times 9 \times 3 = 35$$

$$\frac{4}{\pi} \times 16 \times 4 = 80$$

$$\frac{4}{\pi} \times 25 \times 5 = 160$$

Figure 141. Examples of ratio of presented areas
for Grumman vortex generator.

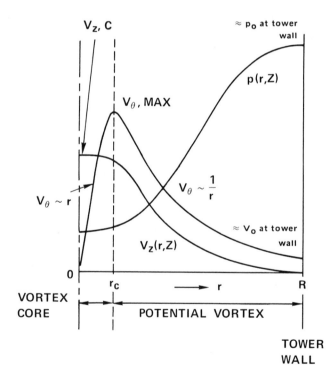

Figure 142. For a closed-botton vortex with nearly uniform V_z in core.

and will reach a maximum in the core of the vortex. He further predicts that the vertical velocity in the core will increase as the diameter of the vortex is increased and, for large-scale systems, could reach values that are at least seven or eight times the angular velocity of the wind at the tower wall (which is, as indicated above, about the speed of the wind entering the tower). Such increases in vertical velocity are indicated in his initial experimental results but need to be verified by tests on larger-scale models and prototypes. On the basis of the above types of analyses, Dr. Yen estimates that large-scale wind energy systems, using ducted turbines interconnected to vortex generators, may be designed with power outputs that are 100 to 1,000 times those of conventional wind turbines of the same diameter, operating at the same ambient free-flow wind speed.

These concepts are expected to provide guidance for further research in developing significant improvements to the design of wind machines of this type in a wide range of sizes; i.e., from a few kW up to perhaps several hundred MW. Because of their enclosed high-speed turbines of relatively small diameters for a given output power, they might be much more suitable for siting in an urban area than would the large-diameter, horizontal-axis wind turbine generators that are designed to operate in ambient windstreams.

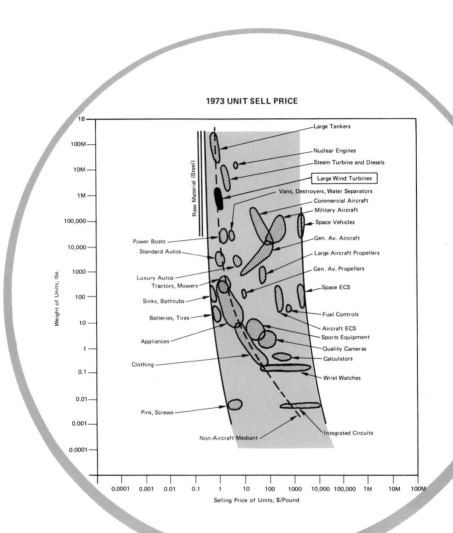

144

8. SYSTEM DESIGN

SYSTEM DESIGN OBJECTIVES

The general objective in designing a WECS is to adequately match the WECS capabilities to the load requirements of the consumer, at a minimum cost of the system to the consumer.

In order to accomplish this, the designer will need to answer the following types of questions about the system.

Power Requirements

- What type of power is required?
- When is the power needed?
- How much power is required at different times?
  —What are the diurnal variations in the power requirements?
  —What are the seasonal variations?
- Is interruptable power sufficient, or is uninterruptable power required?

Wind Availability

- How much wind power density is available at different candidate sites, as a function of time?
- What are the expected variations of wind power density, as a function of height above ground level, at the candidate sites?
- What are the expected diurnal variations in wind power density, at the candidate sites, as a function of season?
- What are the expected seasonal variations in wind power density at the candidate sites?
- How steady is the wind at each candidate site?
- How much turbulence and gusting is expected at each site?
- What is the expected distribution and length of time intervals during which no wind will be blowing, or when the wind velocity will be too high or too low to be used effectively by a WECS, if located at the candidate sites?

Type and Size of WECS Required

- What type of wind machine is best suited for the application planned?
- What is the expected power coefficient of the rotor (or other type of wind energy collector) that will be used?
- What type and amount of supplementary power is required, if any?
- What type and amount of energy storage is required, if any?
- What is the expected power-train efficiency of the WECS, including the gear-train, the electrical generator, or other types of energy converters, if included, and the energy storage unit, if included?
- What energy transmission and distribution losses are expected for the system configuration that has been chosen?
- What size of wind machine is required?
- What system maintenance requirements are expected?

Cost of Energy Delivered

- What are the expected service lifetimes of various components of the WECS?
- What is the expected life-cycle cost of the WECS?
- How will the WECS be financed?
- What interest rates on borrowed capital will be required?
- Will any federal, state or local government economic incentives be available?
- How much usable energy is the WECS expected to deliver over its service lifetime?
- What is the expected average cost of usable energy delivered by the system over its service lifetime?

WECS Viability

- What is the present cost of competing energy?
- How is the cost of competing energy expected to change over the service lifetime of the WECS?
- Can the WECS compete economically with other forms of available energy, over its service lifetime?
- What environmental or sociological factors might contribute to the viability of the WECS?

Background information that should be of assistance in answering these types of questions is contained in this book and in the various documents listed in the appended bibliography.

SYSTEM CHARACTERISTICS

Wind Energy Conversion Systems are primarily characterized by the intermittent nature of the wind, which is their source of energy, and by the fact that their power output varies with both the swept area of their rotors and the cube of the wind velocity (Figures 135 and 136). Because of this cubic relationship between output power and wind velocity, and the fact that the wind is not steady, the actual power output of wind machines, over a long period of time, may be significantly higher than the power output estimated from the average wind speed over that time (see Figure 110). Exact siting is critical, since the wind velocity varies not only with horizontal position but increases with height above the ground level. Local anomalies in wind speed occur because of increased rates of flow over round, smooth hills and other objects and because of decreased rates of flow caused by nearby trees, buildings, and other obstructions.

Furthermore, there are limitations on the size and power output of a WECS because of limitations on strengths of materials, which, in turn, limit the size of blades, bearings, tower heights, and other critical system components. As a result, if a large-capacity WECS is needed for generating electricity for a utility (e.g., 100 MW power capacity, or more), either the system must be composed of a number of interconnected, dispersed WECS units of relatively small size (e.g., up to several MW each) or larger-scale wind powered systems, such as vortex generators, if proven feasible, can be used.

Because of uncertainties in the cost and performance of large-scale WECS, increasing attention has been given in the past few years to the possibility of the widespread use of wind machines in dispersed applications (Figures 143, 144, and 145).[15]

Some of the advantages cited for small, decentralized wind energy systems[31] are listed below.

- They are relatively simple in design.
- Many types of designs have been proven reliable over a long period of years.
- They can be built rapidly, providing energy within months, instead of the years required for large systems.
- Being smaller, more units will be required for a given amount of total output capacity, which would be expected to bring the capital costs of units down more rapidly on the "production learning curve," after mass production is started.
- Generating capacity can be expanded in modular steps to meet changing requirements, thus avoiding the large increments of capacity that are associated with large-scale central power plants.

Figure 143. Dunlite wind-powered generator on roof of rural residence.

- Dispersed wind systems are more labor intensive than are large, central power plants and could therefore help create more job opportunities.
- The technology is exportable; i.e., through the use of dispersed wind energy systems of simple design, it would be feasible for developing nations to greatly increase their energy self-sufficiency.
- Self-sufficiency is also made more feasible for the domestic consumer because the small wind machines used for dispersed applications are relatively easy to assemble, operate, and repair.

The principal arguments for using large, centralized wind energy systems (Figures 146 and 147) appear below.

- They are more compatible with existing utility generation, transmission, and distribution methods.
- They will produce electricity at a lower price to the consumer, since the WECS can be located at sites with high persistent winds and the cost of transmission and distribution of electricity produced at such sites is small enough to make such system configurations cheaper than distributed WECS if the consumer is already connected to a utility network.
- Centralized systems can operate more effectively with energy storage systems of high capacity, such as water-saver systems using

Figure 144. Battery bank with Dunlite wind machine.

large reservoirs, or systems using large natural caverns or aquifers for storage of compressed air.

In any case, if either a constant or load-matching source of power is required, as in the case of electric utilities, wind energy systems must be used in a fuel-saver or water-saver mode of operation (i.e., in parallel with other energy sources) or some suitable means of energy storage must be provided.[17] Because of the intermittent nature of the wind and its diurnal and seasonal variations in many locations, the type of energy storage provided will need to be of high enough capacity and of low enough cost to be able to economically replace standby equipment that would be used in the fuel-saver or water-saver modes of operation.

Because of the cost of both standby equipment and high-capacity storage facilities, increasing attention has also been given, in the past few years, to wind-powered applications with less stringent load-matching requirements, or applications with inherent energy storage capabilities, such as the pumping of water for livestock or for irrigation, reservoir storage or desalination purposes; the heating or cooling of water in underground aquifers for use in the controlled heating or cooling of buildings; or for various types of agricultural or industrial process heat applications,

Figure 145. Jacobs wind-powered generator on roof of New York City tenement building.

such as grinding of grain or other materials, crop drying, etc.; or for automatic processes for the manufacture of organic or inorganic chemicals or fuels, where constant output is not required and the wind-powered system is still effective if operated only when the wind is blowing.[9,15,17]

SYSTEM REQUIREMENTS

In general, therefore, WECS applications can be categorized in terms of the type and amount of output power required, the requirements of the WECS energy consumer for interruptable and uninterruptable supplies of power, and the daily and seasonal variations in these requirements. In many of the applications discussed above, the power supply can be variable or can be interrupted for short periods of time without disrupting the process that is utilizing the power. If the consumer does require an unin-

Figure 146. Saab-Scania wind-powered generator.

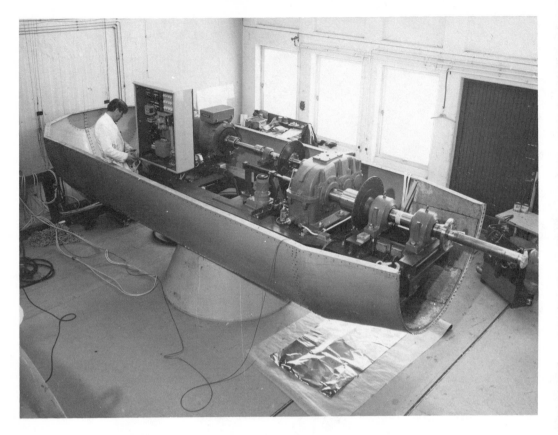

Figure 147. Interior of nacelle of Saab-Scania wind-powered generator.

terrupted source of power, the WECS must be interconnected with a standby source of power (e.g., a motor generator unit or a conventional public utility network) or the WECS must be provided with a means for storing its output energy (e.g., batteries, pumped hydro storage, water-saver storage, compressed air storage, flywheels, or stored electrolytic hydrogen for use in regenerating electricity) over a long enough period of time to cover any slack periods in wind availability (see Figure 95).

EVALUATION CRITERIA

In most WECS applications, the objective will be to design the system to minimize the life-cycle cost of the system and the price of the energy produced by the system. This will require that the capital cost of the WECS, including any energy storage facilities required, be minimized for a specified energy output capacity and that the system be designed to minimize operations and maintenance costs during its complete service lifetime by adequate design of the system components—e.g., blades, generator, bearings, gears, tower, and storage facilities.

Other criteria that should be used in determining the viability of a WECS include:

- The energy playback time, i.e., the time required for the WECS to generate sufficient energy from the wind to equal the amount of energy expended in manufacturing the WECS, as well as operating and maintaining it during this energy payback period);
- The energy gain of the system (i.e., the amount of energy generated by the WECS during its lifetime, divided by the amount of energy required to manufacture, operate, and maintain it during its lifetime); and
- Various possible environmental, aesthetic, legal, financial, institutional, or other types of constraints that might have an impact on the public acceptability of the WECS or its acceptability to public utilities, industries, farm owners, home owners, or other possible users.

In this vein, studies have indicated that the energy payback time for most WECS designs is a few months, and the energy gain of WECS has been estimated at about 40; i.e., about 40 times as much energy will be generated by a WECS during its service lifetime than will be used in manufacturing it. In these regards, WECS compare very favorably with other types of solar—as well as conventional—energy systems.[32]

Likewise, WECS appear to cause very little environmental degradation. The towers and large rotating blades may not be aesthetic in the minds of some people, and there may be some dangers from blade breakage or shedding of ice in adverse weather conditions, depending on the system design.[3] The only other significant environmental impact is the possibility of TV picture and AM radio interference if receivers are located within a few tens of diameters of the rotor blades. Even this potential TV and radio interference is not expected to be a significant problem in areas where cable TV systems are utilized.[4]

COST OF ENERGY PRODUCED

The minimization of the cost of the energy produced by a WECS of a specified rated energy output capacity was studied in the mid-1950's by E. W. Golding (see *The Generation of Electricity by Wind Power*, Philosophical Library, New York, 1956)[8] and later by MITRE/Metrek, General Electric, Kaman Aerospace, Lockheed, Enertech, Parks and Schwind, Clews, and others (see the selected bibliography in the Appendix).

Important parameters in determining the cost of energy produced include the following.

- Wind availability at the selected site measured at some reference height (e.g., 30 feet above ground level) in terms of the annual average wind speed, or preferably in terms of the annual average velocity duration curves for that site, and including estimates of diurnal and seasonal variations in wind speed.

- Estimates of the expected service lifetime of the WECS.
- Height above ground level and/or nearby obstructions, such as buildings, trees, etc.
- Capital costs of the WECS, including those of any energy storage or inverter units utilized by the WECS.
- Rate structure for utility backup energy utilized, or cost of diesel generator or any other backup units utilized by the WECS.
- Installation, operations, and maintenance costs for the WECS.
- Interest rates on borrowed capital used to purchase the WECS.
- Economic incentives provided by federal, state and/or local governments, or other sources.

Capital costs are discussed here in terms of cost per rated kW of output power of the WECS, but could also be examined in terms of the cost per rated hp or the cost per Btu/hour, depending on the WECS application.

It is often preferable to discuss capital costs in terms of the average power output delivered by the system. The ratio of the average power output of an energy system to its rated power output is known as the system "capacity factor" (Figure 148). Since wind machines may be rated at different wind speeds it may be necessary to normalize their rated output power, through the use of their estimated capacity factor for the selected site, or by other means, in order to obtain useful comparisons of capital costs per unit of output power.

Even though present costs of wind machines can be obtained from manufacturers and commercial outlets (Figures 149 and 150), there are still large uncertainties in the ultimate costs of wind machines, energy storage units and other WECS components that are currently under de-

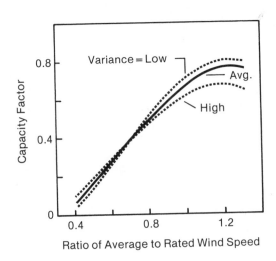

Figure 148. Typical capacity factors for wind machines in various wind regimes.

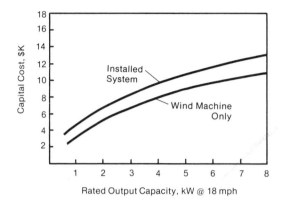

Figure 149. Typical pre-massproduction capital costs of small conventional wind machines.

velopment by government and industry, primarily because of uncertainties in the so-called "learning curves" (Figure 151) which are used to estimate the cost reductions expected to be achieved by mass-production, mass-distribution and mass-installation of wind machines and other WECS components still either under development or in the early stages of production.[27,33]

Usually, the cost of materials will set a lower limit on the ultimate costs of WECS. Recent studies have indicated that the cost of the materials used to build a WECS will be less than $300 per rated kW (rated for winds of 18 mph).[32] However, research and development costs, labor costs, administrative costs, and other manufacturing, distribution, and installation costs are large enough to raise the estimated installed costs of

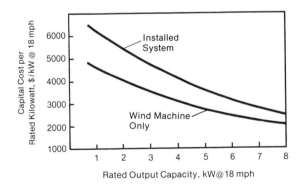

Figure 150. Typical pre-massproduction capital cost, per rated kW, for small conventional wind machines.

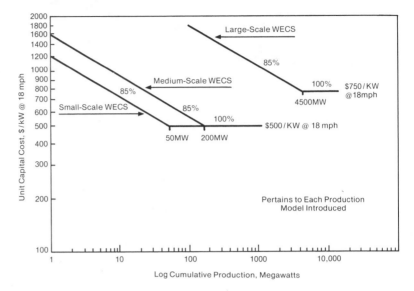

Figure 151. Estimated typical WECS learning curves, in 1978 dollars.

initial production units to levels that are five or ten times the cost of materials alone. Many of the small wind machines that are now commercially available are still being produced with hand labor, and have even higher ratios of installed cost to materials costs than the large-scale machines. The same holds true for some of the types of energy storage units that are under development, particularly those types that do not use aquifers, depleted gas wells, caverns, or other large-scale natural structures with energy storage potential.

In the case of the Tvind wind machine, the installed costs have been kept low through the use of used parts (the gear box and the generator) or components that were mass-produced for other purposes (the bearings), and through the use of student labor during construction. As a result, the installed cost for this machine has been only about $400 per rated kW (rated for winds of 23 mph), with about one-half of this cost being attributed to labor, administrative, and other costs.[18]

Another interesting case in point is the production cost history of water-pumping. multi-bladed, farm-type windmills. It is easy to trace the cost history of these types of wind machines through the Sears Roebuck catalogs. These catalogs are available for each year, from about 1900 to the present, at the U.S. Patent Office Library in Arlington, Virginia, as well as from various publishers of the catalogs. Information on costs of presently available commercial machines of this type is available from the manufacturers (see the Appendix for suppliers). At the turn of the century, when these types of units were being mass-produced (as well as mass-distributed through such means as the Sears Roebuck catalogs), the price was about $300 per rated kW (at 23 mph rated wind speed) for a

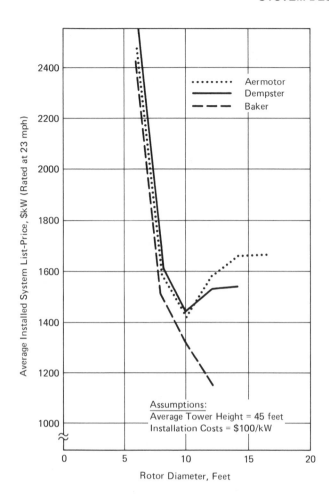

Figure 152. Average installed system list-price, per rated kW, as a function of type of system and rotor diameter.

machine with a 12 foot diameter rotor, installed on a 45 foot tower. This price takes into account the change in the value of the dollar, based on the changes in the consumer price index over that period. This adjusted price held firm until about the mid-1930's, when the rural electrification program was initiated. By 1948, the last year that windmills were listed in the Sears Roebuck catalog, the price had risen to about $500 per rated kW. Today, only about 2,500 of these units are being sold per year. The manufacturing today is done on a piecemeal basis, and the average cost has risen to about $1,500 per rated kW (Figures 152 to 155).[16]

In this regard, it is also interesting to compare the cost of wind machines and solar thermal collectors, on a cost per square foot basis.[15] A 12 foot diameter machine will have about 100 square feet of collector area and will have an output of about 2.5 kW in a 23 mph wind. At $300 per rated kW, the cost will be about $6 or $7 per square foot of swept area.

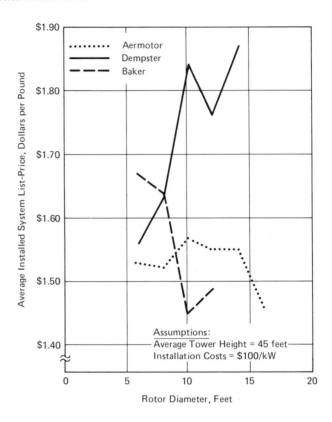

Figure 153. Average installed system list-price, in dollars per pound, as a function of type of system and rotor diameter.

Today, solar thermal collectors, such as those designed to heat water or air-space for households, are selling for about $20 to $25 per square foot of collector area, on the average, which is equivalent to about $1,000 per rated kW (at 23 mph) for a wind machine. The installed cost goal for mass-produced solar collectors is about $10 or $12 per square foot of collector area, which would be equivalent to about $500 per rated kW (at 23 mph) for a wind machine. The installed cost goals for both large and small wind machines that are being developed by the federal government are in the range of $500 to $800 per rated kW (for 18 to 23 mph wind speeds).[2] In comparing the relative costs of energy produced by solar thermal and photovoltaic collectors with that of wind machines, it should be kept in mind that the wind power available in many areas is frequently greater than the direct insolation, on the average, and the wind energy can be converted into a useful form at higher efficiencies with less collector material than for photovoltaic and solar thermal systems. For instance, for modern propellor-type blades, the solidity is only about 1 to 10% while for for photovoltaic and solar thermal collectors, the entire collector area must be covered by the collector.[34]

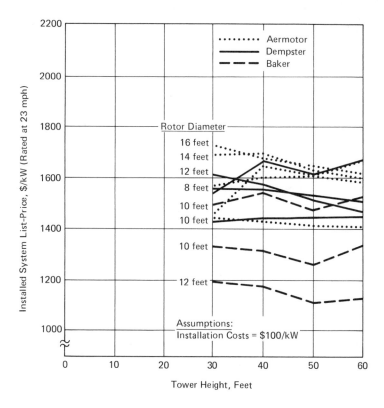

Figure 154. Installed system list-price, per rated kW, as a function of system, rotor diameter, and tower height.

The interconnection costs for large networks of wind machines in centralized utility applications are expected to decrease as the sizes of the wind machines are increased, since lesser numbers of units will be needed for a given output capacity of a network. Likewise, the maintenance costs of a network of wind machines may decrease as the size of the units is increased, since there would be fewer machines to service.

There may be advantages to locating large numbers of wind machines in remote areas where wind speeds are high, and transmitting the output energy to population centers, in the form of electricity, or eventually in the form of a fuel gas such as hydrogen or methane. The economics of this type of system configuration will depend on the relative amounts of wind power density available at the remote sites, compared to amounts available at sites closer to population centers; the relative cost of transmitting electrical power and fuel gases over different distances; the availability of water and/or carbon (e.g., in the form of coal) for manufacturing the fuel gas; and other factors.[19]

As indicated previously, in order to minimize the cost of the output power produced, the design of a wind machine should be matched as

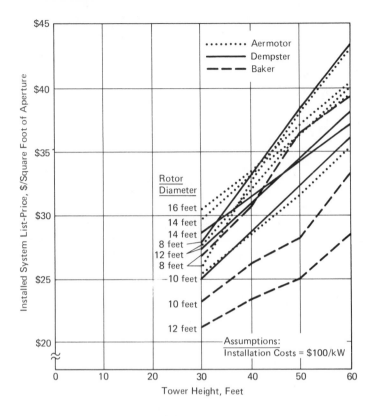

Figure 155. Installed system list-price, per square foot of aperture, as a function of type of system, rotor diameter, and tower height.

closely as possible to the range and relative frequencies of occurrences of the wind speeds in which it is expected to operate, and for the types and variations of loads it is expected to satisfy. In many applications, it may be economical to include energy storage in the system.

STORAGE OPTIONS

If adequate energy storage capacity is added to a wind energy conversion system, it can be operated independently to supply base loads without interconnection to a conventional system. In this case, it can be used as a complete substitute for a conventional electric generating plant.[19]

In other cases, storage can be used with WECS that are interconnected with a utility to displace generated power and to reduce peak loading on the utility system.[20]

Systems capable of storing energy generated from the wind can be categorized in terms of electrochemical energy storage systems (e.g., batteries, or systems that store hydrogen generated by electrolysis), thermal energy storage systems (e.g., systems that store heat produced by high pressure or mechanical motion and friction in a working fluid), kinetic

Table 7. Evaluation of storage technologies.

	Batteries	Hydrogen Storage	Thermal	Fly-wheels	Electro-magnetic	Pumped Hydro	Compressed Air Storage
Minimum Economic Sizes for Utility Application	10 MWh	10 MWh	600 MWh	10 MWh	10,000 MWh	10,000 MWh	100 MWh
Estimated capital costs, $/kWh	18-35	30-40	10-40	40-50	50-60	20-30	20-30
Expected Life, Years	10–20	30	20	30	30	50	30
Dispersed Storage Capability	Yes	Yes	Yes	Yes	No	No	No
Turnaround Efficiency, %	70–80	40–60	High	80	90–95	70	45

Table 8. Incremental increase in busbar price of electricity with energy storage.

$$P_S = \frac{(CC) \times (FCR)}{E \times LF \times 8.76} + \text{O\&M (in mills/kW h)}$$

where,

P_S = Incremental increase in busbar price for energy storage

CC = Capital cost of storage system, $/kW (rated)

FCR = Fixed charge rate (e.g., 0.15)

E = Turnaround efficiency for the storage system

LF = **Capacity factor of the storage system**

O&M = Operations and maintenance costs (e.g., 2-mills/kW h)

energy systems (e.g., flywheels), potential energy systems (e.g., pumped hydro systems or compressed-air systems), and super-conducting magnetic storage systems.

Recent analyses of energy storage systems have evaluated some of these available storage technologies in terms of the minimum economic sizes of energy storage units needed for utility, residential and intermediate (e.g., community or shopping center) applications in different wind regimes, estimated total capital costs (for the energy conversion unit, as well as for the energy-storage container-unit), expected storage system lifetime, the capabilities for dispersed storage, and the estimated turnaround efficiencies of the units (Table 7).[20]

The total capital costs in most of these studies have been based on an energy storage cycle time of about ten hours. This may not be sufficient for some wind energy applications where load interruption cannot be tol-

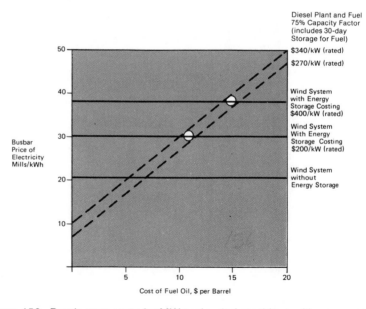

Figure 156. Break-even costs for MW scale wind machines with energy storage.

erated and storage times of tens of days may be required to compensate for short-time slack periods in wind velocities, or where storage times of six months or more may be required to allow for seasonal variations in wind strength. Under these conditions, there may be distinct cost advantages for storage systems (e.g., thermal, hydrogen, or compressed air) that use aquifers, depleted natural gas wells, or other large-scale natural underground structures for the storage of energy generated from the wind. The cost per kWh of storage of such systems is significantly less than that for systems employing man-made structures for energy storage.

If we assume a fixed charge rate of 0.15, a turnaround efficiency of 0.60, a load factor of 0.9, and an O&M cost of 2 mills/kWh, the incremental increase in the consumer price of electricity that would result from the inclusion of energy storage in a WECS would vary from about 9 mills/kWh for a storage system with a capital cost of $200/kW (rated), to about 17 mills/kWh for a storage system with a capital cost of $400/kW (rated) (Table 8 and Figure 156).

In the previous example, where it was estimated that the optimal MW-scale conventional wind system operating in winds with average speeds of 15 mph could produce electricity priced at 20 mills/kWh, the addition of a wind energy storage system costing $200 for each rated kW of the wind machine would increase the busbar cost of the wind-generated electricity to about 30 mills/kWh. If the storage system costs $400/kW (rated), the busbar price would be increased to about 38 mills/kWh. Under these conditions, the break-even point with a diesel plant, operating at a 75% capacity factor and costing between $270 and $390 per rated kW, would occur when the price of fuel oil reached a level of about $11 per barrel in the case of the wind system with energy storage costing $200/kW (rated), and about $15 per barrel for the case with a $400/kW (rated) storage system.

Since both the capital cost per rated kW and the capacity factor of a WECS (which incorporates a given amount of energy storage) increase with the system capacity, the price of energy produced by WECS with storage would not be expected to increase significantly when the capacity of the energy storage unit is increased to provide capacity factors equivalent to those of conventional fossil-fueled or nuclear-fueled systems.

Studies by Justus and others have indicated that widely dispersed settings of WECS units tied into a utility network can increase the amount of firm power produced by the totality of units, and reduce the need for added WECS capacity or increased energy storage to meet a given baseload requirement.[24]

DESIGN OPTIMIZATION

The principal factors that affect the economic viability of wind-powered systems are the availability of wind power at a given site, the investment and O&M costs of the wind-power systems, the expected lifetimes of the systems, the interest rate on investment capital, and the price of competing forms of energy.

The power output, P_o, of a wind machine can be expressed in the following terms:

$$P_o = \frac{(1.35 \times 10^{-6} E)\,(D_r{}^2)(p_b)}{(T + 273.18)}\,(V_r)^3\,\left[C_p(V_r)\right], \text{kW} \tag{1}$$

where

E = efficiency of power train
D_r = diameter of swept area of rotor, meters
p_b = barometric pressure, pascals
V_r = instant wind speed at rotor hub-height above ground level, m/sec
T = ambient temperature, °C
$C_p(V_r)$ = power coefficient for the rotor

$$V_r \quad = V_a \left(\frac{H_r}{H_a}\right)^{1/7}$$

V_a = instant wind speed at anemometer-height above ground level, m/sec
H_r = rotor hub-height above ground level, meters
H_a = anemometer height above ground level, meters

The above algorithm is based on the use of "instantaneous" windspeeds (i.e., where the wind speeds are averaged over a period comparable to the response time of the wind turbine generator). If "long-term wind-speed averages" (i.e., greater than about 1 minute) are used, the energy pattern factor (i.e., $<V^3>/\bar{V}^3$) must be accounted for, where $<V^3>$ is the average of the cube of the instantaneous wind speed and $\bar{V}$ is the long-term wind-speed average. The dotted lines of Figure 157 compare typical energy pattern factors, as a function of mean annual windspeed at 10 meters above ground level, for sites with different amounts of wind-speed variance.[37] Also shown, as a solid line, are estimates of average energy-pattern factors for 26 DOE Solar Meteorological (SOLMET) stations.[39]

The following expression for the mass density of the air, ρ_m, may be derived from the Perfect Gas Law:

$$\rho_m = \frac{p}{RT_a}, \text{kg/m}^3 \tag{2}$$

where

p = absolute pressure of the gaseous system, kg/m²
R = gas constant for air, joules/(kg · °K)
T_a = absolute temperature, °K

Using Equation 2, the mass density at any given temperature and pressure can be estimated. For instance, at 1 atmosphere and 0°C, the mass density of air is:

$$\rho_m = 1.29 \text{ kg/m}^3 \tag{3}$$

A general expression for the mass density of the air, ρ_m, can be obtained in terms of the number of pascals of ambient barometric pressure, p_b, as well as the ambient temperature, T. Substituting these values in the general expression for C_p (p. 129) results in Equation 1.

The coefficient of power, C_p, for the Mod 0 series rotors that have undergone instrument tests is shown in Figure 158 as a function of Θ at the three-quarter span point, and the rotor-tip-speed-to-wind-speed ratio, λ. C_p can also be given in terms of the windspeed at the rotor-hub height, V_r, since:

$$\lambda = R\Omega/V_r \tag{4}$$

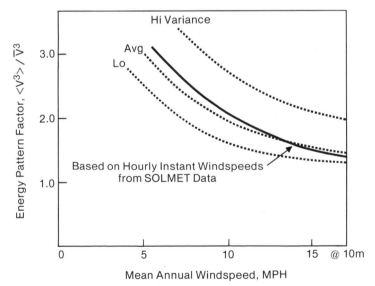

Figure 157. Energy pattern factor as a function of mean annual windspeed and variance.

where

R = radius of rotor-swept area, meters
Ω = angular speed of the rotor, radians/sec

In using the algorithm shown in Equation 1, to calculate the power output of the wind machine, an empirical table can be used that is based on the envelope of the curves for different values of Θ, as shown in Figure 158, and for a constant Ω representing the synchronous drive speed. A cut-in wind speed, V_{in}, and a cut-out wind speed, V_{out}, can be specified, below which and above which, respectively, the power coefficient is assumed to be equal to zero (i.e., representing conditions under which the blade does not rotate).

Pre-massproduction capital costs of wind-powered systems, large enough, for instance, to supply most of the energy for a typical home in a favorable wind regime, are typically in the range of $5000 to $15,000 or $1000 to $3000 per kilowatt when rated in winds of 18 mph. However, many of the new types of WECS that are under development are expected to be mass-produced and mass-distributed in the future, which should reduce the prices of most such systems by a factor of two or three, compared to the pre-massproduction prices.

The assumed WECS production learning curves shown in Figure 151 are based on an extensive survey of the present and projected prices of small-, medium-, and large-scale WECS.[39] The expected minimum prices are a function of the cost of the materials used to manufacture the WECS.

Figures 159 and 160 describe the wind-system design synthesis and the system-selection methodology that were used for the WECS commercialization study described in Reference 39.

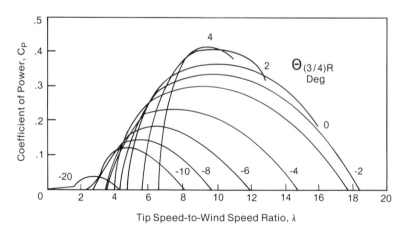

Figure 158.
Power coefficient for Mod OX wind machine.

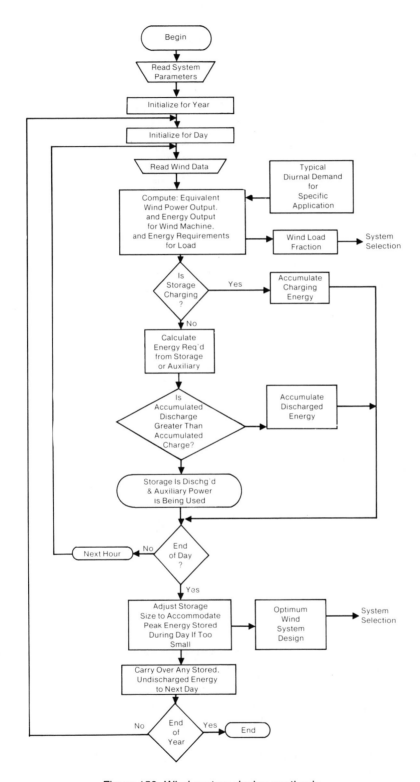

Figure 159. Wind-system design synthesis.

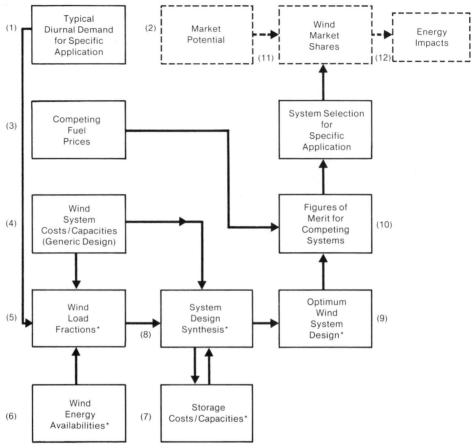

Figure 160. System selection methodology.

Figure 161 shows the diurnal electrical load required for a typical food-processing plant located in Madison, Wisconsin. This is compared with the estimated availability (on two different days) of energy produced by a WECS complex consisting of three Mod OA wind machines located in the Madison area.

Table 9 shows the "wind fraction" (i.e., the fraction of the daily load that would be supplied by wind energy) on these two days, as a function of the rated output capacities of the generators and the number of wind machines utilized by the WECS. In each case, shown in Table 9, it was

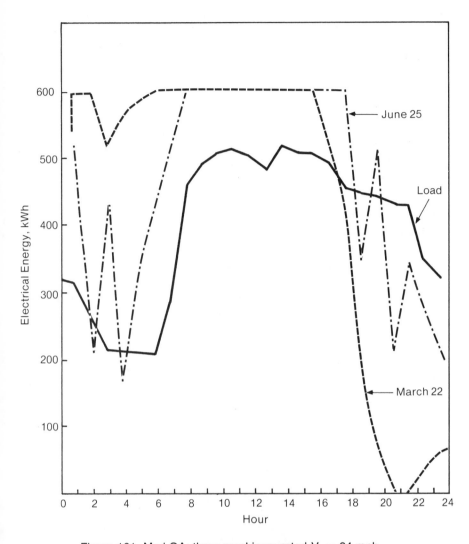

Figure 161. Mod OA; three machines; rated V_r = 24 mph.
Comparison of WECS energy availability and
load requirements for food processing plant, Madison, Wisconsin.

Table 9. Wind fraction, Madison, Wisconsin. No energy storage.

	March 22 No. of Machines			June 25 No. of Machines		
	1	2	3	1	2	3
MOD O	20%	40%	56%	23%	47%	68%
MOD OA	37%	55%	78%	38%	76%	92%
MOD OB	40%	69%	78%	55%	85%	92%

assumed that no energy storage was utilized by the WECS. The following rated capacities were assumed for the generators in the cases shown in Table 9:

- Mod O, 100 kW
- Mod OA, 200 kW
- Mod OB, 400 kW

In each of these cases, the swept diameter of the rotor was assumed to be 125 feet.

Note that the wind fraction increases nonlinearly in the upper ranges, as either the rated capacities of the generators or the number of wind machines is increased. This is illustrated in more detail in Figure 162, which also shows the effect of different amounts of energy storage on the size of the wind fraction. In the particular case shown in Figure 162, it was assumed that systems with small numbers of machines (i.e., for which the system output did not exceed the minimum diurnal load) would be operated in a fuel-saving mode with no energy storage. Above

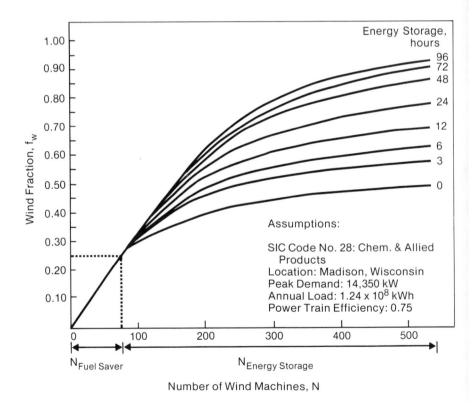

Figure 162. F-chart for Mod OX industrial electric application.

this size, it was assumed that various amounts of energy storage would be used. This affects the wind fraction of the system, as indicated in Figure 162.

ECONOMIC VIABILITY

Figure 163 shows the system capital costs for the case shown in Figure 162, as a function of wind fraction, the number of wind machines used in the WECS, and the number of hours of storage employed, based on the peak load for the system. Note that for each WECS-complex size, i.e., depending on the number of wind machines utilized in the WECS complex, there is a minimum system cost as a function of the amount of energy storage used for the WECS. The dotted line in Figure 163 shows the locus of these minima as a function of wind fraction.

Figure 164 shows the minimum cost of energy delivered by the WECS, as a function of wind fraction, based on the locus of minima shown in Figure 163. Figure 164 also shows the optimum amount of energy storage, for this case, as a function of wind fraction. For instance,

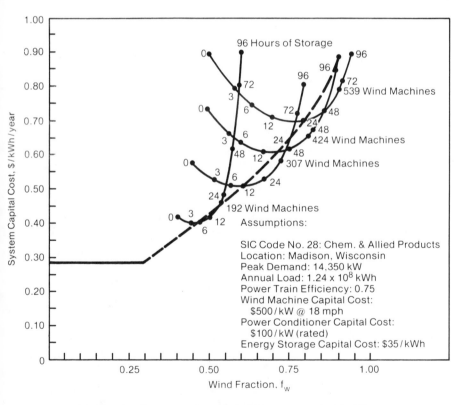

Figure 163. System unit capital costs versus wind fraction

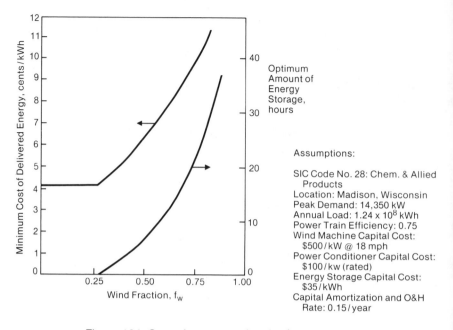

Figure 164. Cost of energy and optimal storage versus
wind fraction.

for the case shown, a WECS that could deliver electricity at $0.06 per kWh would provide a wind fraction of about 0.50 and would require about 5 hours of storage.

Figure 165 illustrates the importance of using all of the energy produced by the WECS for satisfying primary and secondary loads, or feeding the surplus WECS energy back to a utility grid at a fair and reasonable rate. In the case shown in Figure 165, the WECS-generated electricity would compete with electricity, provided at the current average United States utility rate, for all but the high ranges of wind fractions. At lower rates for fed-back power, such WECS would compete effectively only when operated in a fuel-saver mode with no energy storage.

Figure 98A shows the estimated geographical distribution of mean annual wind-power at 50 meters above ground level, for exposed areas of the contiguous United States.[38] Table 10 provides the wind-power equivalency, in terms of the costs of electricity and heat, based on the results of studies described in Reference 39. For instance, for locations with mean annual wind speeds of, say, 400 watts per square meter, as shown in Figure 98A, the ultimate cost of electricity for mass-produced WECS is estimated to be $0.03 per kWh for small-sized machines, $0.02 per kWh for medium-sized, and $0.02 per kWh for large-sized, as shown in Table 10. Corresponding costs of heat would be $8 per

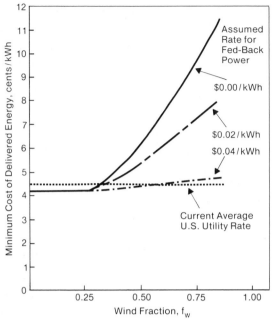

Figure 165. Cost of energy as influenced by sale of
surplus energy.

Table 10. Wind-power equivalency for ultimate small-, medium-, and large-scale WECS.

Wind Power at 50 Meters	Electricity Cost (per kWh)			Heat Cost (per MMBtu)		
	Small WECS	Medium WECS	Large WECS	Small WECS	Medium WECS	Large WECS
100 W/M^2	$0.10	$0.08	$0.09	$30	$23	$25
200	$0.05	$0.04	$0.04	$15	$12	$13
300	$0.03	$0.03	$0.03	$10	$ 8	$ 8
400	$0.03	$0.02	$0.02	$ 8	$ 6	$ 6
500	$0.02	$0.02	$0.02	$ 6	$ 5	$ 5
600	$0.02	$0.01	$0.01	$ 5	$ 4	$ 4
700	$0.01	$0.01	$0.01	$ 4	$ 3	$ 4
800	$0.01	$0.01	$0.01	$ 4	$ 3	$ 3

MMBtu for small-sized machines and $6 per MMBtu for medium- and large-sized machines. Tables 11 and 12 show typical recent prices of utility electricity, and consumer costs for heating oil, in different parts of the country. These can be used with the types of equivalency costs, shown in Table 12, for comparison purposes.

Figure 166 shows the estimated midterm (i.e., mid-80s) capital cost, per unit capacity, for energy produced by an 8-kW SWECS, rated at 18 mph, with a swept diameter of 10 meters, at each of 26 SOLMET Stations, for the locations indicated.[39]

Finally, Table 13 shows, for different locations, estimates of the simple payback period, i.e. the ratio of the capital cost per total energy produced (as shown in Figure 166), divided by the estimated utility price for electricity (as shown for some examples in Table 11). In Table 13 for the near-term case, current federal incentives for residential applications of wind machines (i.e., 30% for the first $2000 and 20% for the next $8000 of capital cost) were included in estimating the capital cost of the SWECS. For the mid-term case, it was assumed that the time limit for such incentives had expired, and the allowances for such incentives were therefore not included.

The general conclusions of the study described in Reference 39 are as follows:

1. *Economic viability of wind-powered fuel-saver systems.* The fuel-saver (i.e., zero energy storage) mode of WECS operation for dis-

Table 11. Typical utility prices for electricity, November 1978.

Boston, MA	$0.059/kWh
New York, NY	$0.085
Scranton, PA	$0.049
Arlington, VA	$0.049
Birmingham, AL	$0.044
Miami, FL	$0.055
Detroit, MI	$0.055
Chicago, IL	$0.054
Madison, WI	$0.029
Wichita, KS	$0.048
Oklahoma City, OK	$0.034
Omaha, NB	$0.033
Denver, CO	$0.044
Salt Lake City, UT	$0.060
Los Angeles, CA	$0.049
Seattle, WA	$0.015
U.S. Average	$0.044

Present escalation rate, U.S. average: 4.5%/year

Table 12. Typical consumer costs for heating oil, January 1979.

Boston, Ma	$3.91/MMBTU
New York, NY	$3.68
Scranton, PA	$3.89
Arlington, VA	$3.80
Birmingham, AL	$4.06
Miami, FL	$4.00
Detroit, MI	$3.91
Chicago, IL	$3.97
Madison, WI	$3.73
Wichita, KS	$3.62
Oklahoma City, OK	$3.60
Omaha, NB	$3.79
Denver, CO	-
Salt Lake City, UT	-
Los Angeles, CA	$4.21
Seattle, WA	-

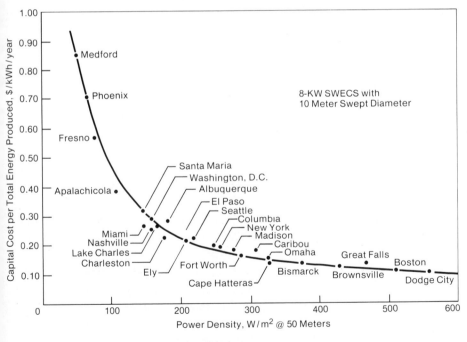

Figure 166.
Capital cost, per unit capacity, as a function
of annual average wind-power density.

Table 13. Simple payback period, SWECS for residential heat pump and electricity, fuel escalation at 2% per year, no-waste SWECS electricity.

Location	Near Term	Mid Term
Ilio Point, Molokai, HI	0.8 years	0.3 years
Kahua Ranch, Hawaii, HI	1.0	0.4
Makawehi, Kauai, HI	1.2	0.5
Kahuku, Oahu, HI	2.4	1.0
Boston, MA	3.7	1.6
New York, NY	4.2	1.8
Dodge City, KS	4.5	2.0
Brownsville, TX	4.5	2.0
Cape Hatteras, NC	5.4	2.3
Bismarck, ND	5.8	2.5
Fort Worth, TX	7.8	3.4
Columbia, MO	8.1	3.5
Caribou, ME	8.1	3.5
Omaha, NB	9.1	4.0
Great Falls, MN	9.3	4.0
Charleston, SC	9.4	4.1
Miami, FL	9.5	4.1
El Paso, TX	9.6	4.2
Albuquerque, NM	9.6	4.2
Madison, WI	11.3	4.9
Santa Maria, CA	12.4	5.4
Ely, NV	15.0	6.5
Lake Charles, LA	15.2	6.6
Washington, D.C.	15.6	6.8
Apalachicola, FL	16.9	7.3
Nashville, TN	17.2	7.5
Fresno, CA	23.8	10.3
Phoenix, AZ	25.4	11.0
Seattle, WA	29.7	12.9
Medford, OR	56.7	24.6

persed generation of industrial onsite electricity, as well as residential and commercial electricity and electrical heat pump, air-conditioning, and hot-water systems, can be expected to be highly competitive, by the midterm, with both conventional and alternative types of energy systems in most areas of the United States. Areas excepted are those characterized by abnormally low wind-power densities and/or low costs of utility-generated electricity. Also by the midterm, the fuel-saver mode of generating industrial process heat should be competitive with other means of generating thermal energy in certain United States regions that possess high average annual wind speeds, i.e. greater than about 12 mph at 10 meters above ground level.

2. *Favorable wind resources and potential markets*. Estimates of the geographical distributions of available wind-power densities in each

state or region of the United States indicate that localized wind-power densities may exceed the geographical average by at least a factor of 2 throughout 10% to 20% of their areas. A corollary to this observation is that no single site investigated can be representative of the wind regime for an entire census region; some locations will prove more favorable—others less desirable—for installing wind machines. The fractional area with sufficient wind-power density to provide WECS break-even costs, currently averaging 4¢ per kWh, should increase in the future, e.g., as fuel prices escalate and/or as the prices of mass-produced WECS decrease. A required step in developing reliable market penetration estimates is to identify geographical areas where high-wind regimes co-exist with high existing or future population and industrial densities.

3. *WECS wind fractions*. WECS used in a fuel-saver (i.e., zero storage) operating mode were found generally to have lower capital costs per unit output energy delivered to the primary load than when energy storage was incorporated. However, higher wind fractions (i.e., the fraction of a specified primary load that is supplied by wind power) can be achieved through the use of energy storage (e.g., through the use of batteries, compressed air, hydrogen, or pumped storage). Also, some energy storage may help to stabilize the WECS outputs in areas having highly variable winds. Thus, the use of some WECS storage capacities may be anticipated, particularly in view of escalating prices of competing energy.

4. *WECS payback periods*. In some areas of the United States (e.g., within Alaska, Hawaii, New England, New York, western Kansas, Texas and Oklahoma panhandles, southern Texas, eastern Wyoming, and along the Atlantic capes of North Carolina) small-scale WECS (SWECS) appear to be commercially viable today, particularly for residential applications, to which federal incentives contained in the National Energy Act of 1978 apply. The number of commercially viable sites for WECS applications is expected to increase rapidly as WECS fuel-saver system prices decline, through mass production, to the order of $500 to $750 per kilowatt, at rated wind speeds of 18 mph, and as the prices of conventional fuels increase. The studies reported herein provide evidence that in about half of the United States, simple payback periods of 5 years or less can be achieved with some horizontal axis WECS generic designs by the time that production price stabilization occurs (i.e., the mid-80s). Here, the simple payback period is defined as the time required to pay back the installed capital cost of the wind machine (without regard to 0&M costs, interest charges on capital investment, or other fixed charges) with savings accrued from the price of conventional energy replaced by those derived from wind energy.

5. *Minimum waste fraction as a market leverage factor*. In applications where all of the WECS output electricity can be used or where excess WECS energy can be sold to a utility at cost, the resulting capital costs per unit of system energy delivered are about half those for systems that discard excess energy as waste. This differential is equivalent to the effect of an 85% cost learning curve applicable over about four doublings of cumulative production or, alternatively, to a 35-year fuel price escalation at 2% per annum. Utility buyback of excess WECS generation at prices greater than about 4¢ per kWh significantly improves the near-term WECS economics. Sharp increases in WECS-generated electrical energy fed to utilities are to be anticipated when the cost of wind-generated energy drops below a price equivalent to the utility's wholesale rates paid for bulk purchases of electric energy, plus the utility's unit costs for providing backup power to the WECS site (i.e., the utility's cost for standby generating capacity and equipment for transmission and distribution).

Several types of economic incentives have been included in the recent national energy legislation that provide income-tax breaks and investment credits for certain types of wind-machine applications. Still other federal, state, and local government incentives, such as loan guarantees, low interest-rate loans, and property tax breaks, can be used to reduce the life-cycle costs of wind-powered systems and stimulate the use of wind-power in the near-term.

One QUAD of energy is equivalent to . . .

Enough energy to heat 500,000 homes for 20 years.*

* Based on an 1,800 square foot, single family, frame house in the Washington D.C. metropolitan area.

Enough crude oil to fill a fleet of 75 supertankers.*

* Based on 325,000-ton supertankers, each with a capacity of 2.3 million barrels.

A mountain of coal 1,000 feet high and 2,000 feet in diameter, or 500,000 railroad cars* full of coal.

* Based on 83 tons per car.

9. FUTURE UTILIZATION

There has been considerable speculation concerning the total amount of wind energy that will be used to help supply energy needs of the future and how and when this wind energy will be utilized. There is general agreement that wind is a very clean, replenishable source of energy and, that though it is intermittent in nature and relatively dilute as compared to fossil and nuclear fuels, it constitutes a very large but practically untapped energy resource. How effectively and to what extent it will be used in the future depends on our energy needs and our ingenuity.

Dr. Marvin Gustavson of the University of California has estimated that about 4,000 quads per year of energy are potentially available, worldwide, in the near-surface winds of the earth, of which about 60 quads per year are available over the land areas of the 48 contiguous states of the United States. Lockheed estimates that by siting wind machines only on open-range lands of the conterminous United States, a total of over 150 quads per year of primary wind energy could be converted to forms useful to man. If the Betz coefficient and the conversion efficiencies of these machines are taken into account, as in the Gustafson estimate, this amounts to about 50 quads per year of usable energy. In the Lockheed definition of "open range land," populated areas, privately owned land, national or state parks, military reservations, scenic reserves, and transportation corridors and rights-of-way are all excluded. Lockheed further estimates that offshore, within 15 miles of the coastlines of the United States, about 70 quads of additional primary wind energy is available, which could be converted into about 20 quads of useful WECS output energy.

On the basis of these estimates, the total amount of useful WECS ouput energy potentially available to the United States is roughly equivalent to the present and foreseeable demands for all types of energy in the United States. A total of about 75 quads per year of primary energy resource is being used in the United States at present, to satisfy end-use energy demands of about 45 quads per year. By the year 2000, it is expected that end-use demands will rise to about 70 quads per year, requiring about 110 quads per year of primary energy resources.

Because of the relative simplicity of wind energy systems compared to energy systems of other types, and because of the worldwide availability of wind resources, it is expected that large markets for WECS will develop in other countries, as well—particularly in the underdeveloped countries of the world.

Even though wind resources are large, there are a number of problems that must be addressed if wind energy is to be used effectively and extensively in the future. New, efficient, reliable designs for wind machines must be developed and proven in a wide variety of potential applications. Manufacturing facilities must be provided and adequate marketing, distribution, installation, operations, and maintenance infrastructures must be developed. In addition, suitable sites must be chosen for the wind machines that are installed. All of this will require time and effort.

Taking these factors into account, a recent MITRE/Metrek study concludes that wind energy, under an accelerated development program (with federal initiatives of the type provided in the National Energy Plan of 1977) might, by the year 2000, be expected to displace about 2 quads per year of fossil fuels used by utilities for the generation of electricity—and about 6 quads by the year 2020.[36] Another recent study, by the Council for Environmental Quality of the Executive Office of the President, indicates that, with an accelerated Federal Wind Energy Program, wind energy might be expected to displace 4 to 8 quads per year of fossil fuels by the year 2000, and 8 to 12 quads per year by the year 2020.[31] Estimates by Lockheed, General Electric,[27] Stanford Research Institute,[34] and others indicate that the fossil fuel displacement by wind energy by the year 2020 might be as high as 30 quads per year.

In summary, while wind machines are inherently among the simplest of energy conversion devices, the wind itself is a diffused and variable medium. The selection of optimal siting for wind machines and the choice of optimum designs for particular applications are often difficult and complex problems. Much research and development work on wind machines still remains to be done, but wind machines appear to have a high potential for making significant contributions to the goal of meeting our future energy needs through the use of clean and essentially inexhaustible sources of energy.

Figure 167. Mounting wind machine on tower at Rocky Flats test site, Colorado.

Figure 168. Testing small-scale wind machines at Rocky Flats, Colorado. Left to right: Sencenbaugh, Jacobs, Zephyr Wind Dynamo, American Wind Turbine and Grumman wind machines.

APPENDIX

a.c.   Alternating current.

Aleutian Arc   Chain of Aleutian Islands.

Ampere   A unit of electrical current.

Anemometer   An instrument for measuring wind speed.

Average wind speed   The mean wind speed over a specified period of time.

Bedplate   A base-plate for supporting a system component or structure.

Break-even costs   The system costs at which the price of a system's product is equal to the price of the equivalent energy product of another type of system.

Busbar price   The price of electricity at a generating plant; does not include price increment resulting from transmission and distribution costs.

Capacity factor   The average power output of an energy system divided by its rated power output.

Capital costs   Investment costs required to build a system or device.

Chord   The distance from the leading to the trailing edge of an airfoil.

Concentrator   A device or structure that concentrates a windstream.

Crosswind   Crosswise to the direction of the windstream.

Cut-in speed   The wind speed at which a wind machine is activated, as the wind speed increases.

d.c.   Direct current.

Diffuser   A device or structure that diffuses a windstream.

DOE   Department of Energy

Downwind   On the opposite side from the direction from which the wind is blowing.

Energy   The capability of doing work.

Energy density   The amount of energy flowing in a windstream divided by the cross-sectional area of the windstream.

ERDA   Energy Research and Development Administration.

Eutectic (salt)   A material with a relatively low melting point and large heat of fusion.

Fantail   A small wind machine mounted on an axis perpendicular to the shaft of a horizontal-axis rotor, and used to keep the larger machine headed into the wind.

Furling speed   The wind speed at which a wind machine is shut down to avoid damage from high winds.

Gigawatt A measure of power, equal to 10^9 watts.

Head The height of water in a reservoir.

hp Horsepower, a measure of power capacity.

Inverter A device that converts d.c. to a.c.

kV Kilovolt, a measure of electrical potential or potential difference, equal to 10^3 volts.

kVA One thousand volt-amperes, a measure of power capacity.

kW Kilowatt, a measure of power, equal to 10^3 watts.

kWh Kilowatt-hour, a measure of energy, equal to 10^3 watt-hours.

Laminar (flow) Smooth.

Lift-type devices Devices that use airfoils or other types of shapes that provide aerodynamic lift in a windstream.

m Meters.

mph Miles per hour.

MW Megawatt, a measure of power, equal to 10^6 watts.

NOAA National Oceanic and Atmospheric Administration.

NASA National Aeronautics and Space Administration.

O&M costs Operations and maintenance costs.

Panemone A vertical-axis wind collector capable of reacting to horizontal winds from any direction.

Peak-shaving The process of supplying power from an extraneous demand on a system.

Pinned truss An assembly of connected beams.

Polder A tract of low land reclaimed from the sea by dikes, dams, etc.

Power The amount of work done per unit time.

Power coefficient The ratio of the power extracted by a wind machine to the power available in a windstream.

Power Coefficient, Cp The ratio of the power delivered by a wind machine to the power available in a windstream.

Power density The amount of power per unit of cross-sectional area of a windstream.

Rated output capacity The output power of a wind machine operating at the constant speed and output power corresponding to the rated wind speed.

Rated wind speed The lowest wind speed at which the rated output power of a wind machine is produced.

REA Rural Electrification Administration.

rpm Revolutions per minute.

Shear A relative motion parallel to the surface of contact.

Shroud A structure used to concentrate or deflect a windstream.

Solidity The ratio of the projected area of a rotor, on a plane perpendicular to its axis of rotation, to the swept area of the rotor.

Stock   A bar used to support a windmill sail.

Synchronous generator   An a.c. generator that is synchronized with an exterior a.c. power source.

Synchronous inverter   A d.c. to a.c. inverter, the output of which is synchronized with an exterior a.c. power source.

Tip speed-to-wind speed   Ratio of the speed of the tip of a propeller blade to the speed of a windstream in which it is located.

Torque   The product of a turning force and the perpendicular distance from the line of action of the force to the axis of rotation.

Translated (motion)   In a straight line.

Turbulence   Non-laminar flow in a fluid.

Turnaround efficiency   The resulting efficiency when energy is converted from one form or state to another form or state, and then reconverted to the original form or state.

Upwind   On the same side as the direction from which the wind is blowing.

Volt   A unit of electrical potential or potential difference.

Watt   A unit of electrical power: watts = volts $\times$ amperes.

WEC   Wind Energy Conversion.

WECS   Wind Energy Conversion System.

Wind rose   The pattern formed by a diagram showing vectors representing wind velocities over a period of time.

Wind turbine generator   A wind machine, powered by a rotating blade or propeller, that drives an electric generator.

SUPPLIERS

Water Pumping Windmills

Commercial Units

**Aeromotor Division,
   Braden Industries**
P.O. Box 1364
Conway, AR 72032

American Wind Turbine Co., Inc.
1016 E. Airport Road
Stillwater, OK 74074

Dempster Industries, Inc.
P.O. Box 848
Beatrice, NB 68310

H. J. Godwin Ltd.
Quenington, Gloucester
England GL7 5BX

Heller Aller Company
Perry and Oakwood St.
Napoleon, OH 43545

K.M.P. Lake Pump Mfg. Co., Inc.
P.O. Box 441
Earth, TX 79031

Sparco
P.O. Box 420
Norwich, VT 05055

Wadler Manufacturing Co., Inc.
Route 2, Box 76
Galena, KS 66739

Wind-Powered Electric Generators

Commercial Units

Aerolectric
13517 Winter Lane
Cresaptown, MD 21502

Aeropower
2398 4th St.
Berkeley, CA 94710

**Aerowatt S.A./Automatic
   Power, Inc.**
37 Rue Chanzy
75-Paris 11º France

Altos, The Alternate Current
Boulder, CO 80302

Amerenalt Corporation
P.O. Box 905
Boulder, CO 80302

American Wind Turbine Co., Inc.
1016 E. Airport Road
Stillwater, OK 74074

Bucknell Engineering
10717 Rush Street
So. El Monte, CA 91733

Dakota Wind & Sun, Ltd.
P.O. Box 1781
Aberdeen, SD 57401

Domenico Sperandio & Ager
Via Cimarosa 13-21
58022 Folloncia (GR)
Italy

Dominion Aluminum Fabricators
3570 Hawkestone Road
Mississauga, Ontario, L5C 2V8
Canada

Dunlite Electrical Products Co.
Div. of PYE
28 Orsmond St.
Hindmarch, S. Australia

Dyna Technology, Inc.
P.O. Box 3263
Sioux City, IA 51102

Dynergy
Laconia, NH 03246

Edmund Scientific Company
380 Edscorp Bldg.
Barrington, NJ 08007

Elecktro GmbH
Winterthur, Schqeiz
St. Gallerstrasse 27
Switzerland

ENAG s.a.
Rue de Pont-l'Abbe
Quimper (Finisterre)
France

Energy Development Co.
179E, RD#2
Hamburg, PA 19526

Energy Research Products, Inc.
508 S. Burne Road
Toledo, OH 43609

Enertech, Inc.
P.O. Box 420
Norwich, VT 05055

Environmental Energy Engineering
27996 Gardenia Drive
North Olmstead, OH 44070

Grumman Energy Systems
4175 Veterans Memorial Highway
Ronkonkoma, NY 11779

Industrial Inst. Ltd.
Stanley Rd., Bromley BR29JF
Kent, England

Jacobs Wind Electric Co.
Route 13, P.O. Box 722
Fort Meyers, FL 33901

Kaman Aerospace
Old Windsor Road
Bloomfield, CT 06002

KEDCO, Inc.
9016 Aviation Blvd.
Inglewood, CA 90301

Lubing
Maschinenfabrik, Ludwig Bening
P.O. Box 171
D-2847 Barstorf, Vasttyskland
Germany

Millville Windmills & Solar Equipment Co.
P.O. Box 32
Millville, CA 96062

Natural Power, Inc.
New Boston, NH 03070

Noah Energy Systems S.A.
Case Postale 81
CH-1211 Geneve 19
Geneva, Switzerland

North Wind Power Co.
P.O. Box 315
Warren, VT 05674

Pinson Energy Corporation
P.O. Box 7
Marston Mills, MA 02648

Product Development Institute
508 S. Byrne Road
Toledo, OH 43609

SAAB-SCANIA
S-581 88 Linkoping
Sweden

F. L. Schmidt Co.
69 Skt Klemensuzj
Hjallese, Denmark 5260

Sencenbaugh Wind Electric
P.O. Box 11174
Palo Alto, CA 94306

Tvind College
DK-6990 Ulfborg,
Denmark

Standard Research, Inc.
P.O. Box 1291
East Lansing, MI 48823

J. Taylor Co.
88 Hull Road
Woodmanseij
East Yorkshire HU 17 OTH
United Kingdom

USSR Enercgomashexport
35 Mosfilmovskaya UI
Moscow V 330
Russia

Whirlwind Power Co.
P.O. Box 18530
Denver, CO 80218

Winco Division of Dyna Technology
P.O. Box 3263
Sioux City, IA 51102

Wind Power Systems, Inc.
P.O. Box 17323
San Diego, CA 92117

Winduser Co.
P.O. Box 925
Hurricane, UT

Windworks
Box 329, Route 3
Mukwonago, WI 53149

Winpower Corp.
1207 1st Avenue E
Newton, IA 50200

Zephyr Wind Dynamo Company
P.O. Box 241
Brunswick, ME 04011

Prototype Units

Boeing Construction
955 L'Enfant Plaza N., S.W.
Washington, D.C. 20024

Grumman Energy Systems
4175 Veterans Memorial Hwy.
Ronkonkoma, NY 11779

General Electric Space Division
P.O. Box 8555
Philadelphia, PA 19101

McDonnell Douglas Aircraft
P.O. Box 516
St. Louis, MO 63166

TWR Enterprises
72 W. Meadow Lane
Sandy, UT 84070

United Technologies Corporation
One Financial Plaza
Hartford, CT 06101

U.S. Windpower Associates
25 Adams Street
Burlington, MA 01803

WTG Energy Systems
P.O. Box 87
1 LaSalle St.
Angola, NY 14006

Wind Power Products Co., Inc.
213 Boeing Field Terminal
Seattle, WA 98108

Wind Energy Systems

Budgen & Assoc.
72 Broadview Ave.
Pointe Claire 710
Quebec, Canada

Edmund Scientific Company
380 Edscorp Bldg.
Barrington, NJ 08007

Energy Alternatives
Box 223
Leverett, MA 01054

Enertech Corporation
P.O. Box 420
Norwich, VT 05055

Environmental Energies, Inc.
21243 Grand River
Detroit, MI 48219

Garden Way Laboratories
Charlotte, VT 05445

Independent Energy Systems
6043 Sterrettania Road
Fairview, PA 16415

Independent Power and Developers
Box 1467
Noxon, MT 59853

Low Impact Technology
73 Molesworth St.
Wadebridge
Cornwall, England

Natural Power, Inc.
New Boston, NH 03070

Northwind Power Co.
P.O. Box 315
Warren, VT 05674

O'Brock Windmill Sales
Route 1, 12th St.
North Benton, OH 44449

Penwalt Automatic Power
213 Hutcheson St.
Houston, TX 77003

Prairie Sun & Wind Company
4408 62nd Street
Lubbock, TX 79414

Quirk's Victory Light Co.
33 Fairweather St.
Bellevue Hill N.S.W.
Australia

REDE Corporation
P.O. Box 212
Providence, RI 02901

Real Gas and Electric Co.
Box A
Guerneville, CA 65446

Sencenbaugh Wind Electric
Box 11174
Palo Alto, CA 94306

Solar Energy Company
810 18th St., N.W.
Washington, D.C. 20006

Solar Wind Co.
R.F.D. 2
East Holden, MN 04429

Wind Engineering Corp.
Airport Industrial Area
P.O. Box 5936
Lubbock, TX 79417

Windlite Alaska
Box 43
Anchorage, AK 99510

Wind Power Systems, Inc.
Post Office Box 17323
San Diego, CA 92121

Plans and Kits

Earth Mind
2651 O'Josel Drive
Saugus, CA 91350

Enertech
P.O. Box 420
Norwich, VT 05055

Flanagan's Plans
2032 23rd St.
Astoria, NY 11105

Forrestal Campus Library
Princeton University
Princeton, NJ 08540

Homecraft
2350 W. 47th St.
Denver, CO 80211

Jack Park
Box 4301
Sylmar, CA 91342

Sencenbaugh Wind Electric
P.O. Box 11174
Palo Alto, CA 94306

Total Environmental Action, Inc.
Church Hill
Harrisville, NH 03450

Windy Ten
Box 111
Shelby, MI 49445

Windworks, Inc.
Route 3, Box 44A
Mukwonago, WI 53149

Components

Towers

Advance Industries
2301 Bridgeport Drive
Sioux City, Iowa 51102

Eldon Arms
Box 7
Woodman, Wisconson 53827

Natural Power, Inc.
New Boston, NH 03070

Northwind Power, Inc.
P.O. Box 315
Warren, VT 05674

Rohn Manufacturing Co.
Unarco-Rohn, Inc., P.O. Box 2000
Peoria, Ill. 61601

SAC Company
Box 916-T
Parsons, Kansas 67357

Wind Turbine Blades

American Wind Turbine, Inc.
1016 East Airport Road
Stillwater, OK 74074

Gurnard Mfg. Corp.
100 Airport Road
Beverly, MA 01915

Kaman Aerospace
Old Windsor Road
Bloomfield, CT 06002

Senich Corp.
Box 1168
Lancaster, PA 7609

Synchronous Inverters

Windworks
Box 329, Route 3
Mukwonago, WI 53149

Non-Synchronous Inverters

Amerenault
P.O. Box 905
Boulder, CO 80302

ATR Electronics
St. Paul, Minn.

Carter Motor Co.
Chicago, Il.

Dynamote Corp.
Seattle, Washington

Elgar Corp.
San Diego, CA

Nova Electric Mfg. Co.
Nutley, NJ 07110

Soleq Corp.
Chicago, Ill.

Topaz Electronics
San Diego, CA

Tripp-lite Div. of Trippe Mfg. Co.
Chicago, Ill. 60606

Wind Power Systems, Inc.
Post Office Box 17323
San Diego, CA 92121

Batteries

C&D Batteries Division
Plymouth Meeting, PA

Exide Batteries
ESB Brands, Inc.
P.O. Box 6949
Cleveland, Ohio 44101

Globe Union Inc.
Milwaukee, Wisconsin

Gould Inc.
Industrial Battery Division
Langhorne, PA

Keystone Battery Corp.
Winchester, MA

Surrette Storage Battery Co.
Salem, MA

Trojan Batteries
San Francisco, CA

Wisco Division ESB Inc.
Raleigh, NC

Mule Battery Co.
Providence, RI

Gates Batteries
Gates Energy Products
Gates Tire and Rubber Co.
South Broadway
Englewood, Colorado

Controls

Amerenault
P.O. Box 905
Boulder, CO 80302

Natural Power, Inc.
New Boston, NH 03070

North Wind Power Co.
P.O. Box 315
Warren, VT 05674

West Wind
c/o Geoffrey Gerhard
3095 W. Boyd Drive
Farmington, NM 87401

Generator Units

Dunlite Division, Pye Indutries
Huntingdale, Victoria, Australia

Georator Corporation
Manassas, VA

Zephyr Wind Dynamo Co.
Brunswick, Maine

Wind Measurements

Anemometers

Aircraft Components
North Shore Drive
Benton Harbor, MI 49022

**Bendix Environmental
  Science Division**
1400 Taylor Avenue
Baltimore, MD 21204

Climatronics Corporation
1324 Motor Parkway
Hauppauge, NY

Climet Instruments Co.
1620 West Colton Avenue

P.O. Box 1165
Redlands, CA 92373

Danforth
Division of the Eastern Co.
Portland, MN 04103

Davis Instrument Mfg. Co., Inc.
513 East 36th St.
Baltimore, MD 21218

Dwyer Istruments, Inc.
P.O. Box 373
Michigan City, IN 46360

Electric Speed Indicator Co.
12234 Triskett Road
Cleveland, OH

Enertech Corporation
P.O. Box 420
Norwich. VT 05055

Helion
P.O. Box 445
Brownsville, CA 95919

Kahl Scientific Instrument Corp.
Box 1166
San Diego, CA 92022

Maximum, Inc.
8 Sterling Drive
Dover, MA 02030

Meteorology Research Inc.
Box 637
Altadena, CA 91001

Natural Power, Inc.
New Boston, NH 03070

North Wind Power Co.
P.O. Box 315
Warren, VT 05674

Taylor Instruments
Arden, NC 28704

Texas Electronics Inc.
5529 Redfield Street
P.O. Box 7151 Inwood Station
Dallas, TX 75209

Weather Measure Corporation
P.O. Box 41257
Sacramento, CA

Robert E. White Instruments, Inc.
33 Commercial Wharf
Boston, MA 02110

Wind Power Systems, Inc.
Post Office Box 17323
San Diego, CA 92121

WTG Energy Systems, Inc.
P.O. Box 87
1 LaSalle St.
Angola, NY 14006

R. M. Young Company
42 Enterprise Drive
Ann Arbor, MI

Meteorological Services

**Bendix Corp., Environmental
  Science Division**
1410 Taylor Avenue
Baltimore, MD 21204

Enertech
P.O. Box 420
Norwich, VT 05055

**Meterological Research,
  Inc.(MRI)**
Altadena, CA

The Research Center (TRC)
125 Silas Deane Highway
Wethersfield, CT

REFERENCES

1. Gustavson, M.R., "Limits to Wind Power Utilization," Science, Volume 204, pp. 13–17, 6 April 1979.

2. United States Department of Energy, "Federal Wind Energy Program, Program Summary," January 1978.

3. Rogers, S., et al., "An Evaluation of the Potential Environmental Effects of Wind Energy Systems Development," Battelle Memorial Institute, August 1976.

4. Senior, T.B.A., et al, "Radiation Laboratory TV and FM Interference by Windmills," University of Michigan, February 1977.

5. Mayo, L.H., et al., "Legal-Institutional Implications of Wind Energy Conversion Systems," George Washington University, September 1977.

6. Putnam, P.C., "Power from the Wind," Van Nostrand Reinhold, New York, 1948.

7. Private Communication, "On the Number of Wind Machines in Denmark from 1900 to 1950." P.L. Olgaard, Technical University of Denmark, February 1978.

8. Golding, E.W., "The Generation of Electricity by Wind Power," Philosophical Library, New York, 1956.

9. Eldridge, F.R., ed., "Wind Energy Conversion Systems, Proceedings of the Second Workshop," The MITRE Corporation, October 1975.

10. Hutter, U., "Operating Experience Obtained with a 100-kW Wind Power Plant," Kanner Associates, 1964.

11. Maughmer, M.D., "Optimization and Characteristics of a Sailwing Windmill Rotor," Princeton University, March 1976.

12. Yen, J.T., "Tornado-Type Wind Energy System," Tenth Intersociety Energy Conversion Engineering Conference, August 1975.

13. Wetherholt, L., ed., "Vertical Axis Wind Turbine Technology Workshop," Sandia Laboratories, May 1976.

14. Thomas, R.L., "Large Experimental Wind Turbines—Where We Are Now," NASA–Lewis Laboratories, March 1976.

15. Eldridge, F.R., "Commercialization of Small-Scale Wind Machines," The MITRE Corporation, M77-16, Presentation to the U.S. House of Representatives Ways and Means Committee, June 1977.

16. Eldridge, F.R., "A Survey of Small-Scale, Water-Pumping Wind Machines," The MITRE Corporation, August 1977.

17. Kornreich, T.R., ed., "Third Wind Energy Workshop," JBF Scientific Corporation, March 1978.

18. Tvind Schools, "The Wind Machine at Tvind," Ingenioren, No. 26, June 1976, Reprint Available from MITRE/Metrek, McLean, VA 22101.

19. Eldridge, F.R., "Wind Energy Conversion Systems Using Compressed Air Storage," The MITRE Corporation, M76-39, July 1976.

20. General Electric Company, "Applied Research on Energy Storage and Conversion for Photovoltaic and Wind Energy Systems," January 1978.

21. Heronemus, W.E., "Investigation of the Feasibility of Using Windpower for Space Heating in Colder Climates," University of Massachusetts, 1975.

22. BHRA Fluid Engineering, "Proceedings of the International Symposium on Wind Energy Systems," St. John's College, Cambridge, September 1976.

23. Private communication, Hans Meyer, Windworks, Mukwonago. Wisconsin, March 1978.

24. Justus, C.G., "Wind Energy Statistics for Large Arrays of Wind Turbines," Georgia Institute of Technology, August 1976.

25. Ramsdell, J.V., "Annual Report of the Wind Characteristics Program Element," Battelle-Pacific Northwest Laboratories, June 1977.

26. Reed, J.W., "Wind Power Climatology of the United States," Sandia Laboratories, May 1975.

27. General Electric, Space Division, "Wind Energy Mission Analyses," February 1977.

28. Reed, J.W., "Wind Power Climatology," Weatherwise, December 1974.

29. Ljungstrom, O. and Sodergard, B., "Wind Power in Sweden—A Preliminary Feasibility Study," Swedish Board for Technical Development, May 1974.

30. Wilson, R.E. and Lissaman, P.B.S., "Applied Aerodynamics of Wind Power Machines," Oregon State University, May 1974.

31. Council on Environmental Quality, "Solar Energy—Progress and Promise," Executive Office of the President, April 1978.

32. Institute for Energy Analysis, "Net Energy Analysis of Five Energy Systems," Oak Ridge Associated Universities, September 1977.

33. Coty, U. and Vaughn, L., "Effects of Initial Production Quantity and Incentives on the Cost of Wind Energy," Lockheed-California Co., January 1977.

34. SRI-International, "Solar Energy Research and Development—Program Balance," February 1978.

35. Lockheed-California Company, "Wind Energy Mission Analysis," October 1976.

36. Bennington, G., et al., "Solar Energy—A Comparative Analysis to the Year 2020," MITRE Corporation/Metrek Division, MTR-7579, March 1978.

37. Justus, C.G., "Winds and Wind System Performance," The Franklin Institute Press, 1978.

38. Elliott, Dennis L., "Synthesis of National Wind Energy Assessments," Battelle Pacific Northwest Laboratories, BNWL-2220, WIND-5, July 1977.

39. Eldridge, Frank R. and W.E. Jacobsen, "Distributed Windpower Systems," The MITRE Corporation, MTR-79W00021, May 15, 1979.

SELECTED
BIBLIOGRAPHY

GENERAL REPORTS

Abbott, I.H. and E.E. Von Doenhoff, "Theory of Wing Sections," 1959.

Baumeister, T. and L.S. Marks, eds., "Standard Handbook for Mechanical Engineers," 7th Edition, McGraw-Hill Book Co., 1967.

BHRA Fluid Engineering, "Proceedings of the International Symposium on Wind Energy Systems," St. Johns College, Cambridge, U.K., September 7-9, 1976, Available from BHRA, Cranfield, Bedford, England.

California, University of, "Wind, Waves, and Tides," Marshal F. Merriam, Berkeley, California, NBI-77-39.

Climatic Atlas of the U.S., 1968.

Colorado State University, "Energy from the Wind: Annotated Bibliography," B.L. Berke and R.N. Meroney, August 1975.

Dennis, Laudt and Lisl, "Catch the Wind," Four Winds Press, 50 West 44th St., New York, NY 10036, 1976.

Earthwind, "Wind and Windspinners," M.A. Hackleman, 26510 Josel Dr., Saugus, CA 91350, 1974, 115 pages.

Energy Research and Development Administration, Division of Solar Energy, "Development of an 8 kW Wind Turbine Generator," Federal Wind Energy Program; Request for Proposal No. PF64086F: March 28, 1977.

Energy Research and Development Administration, Division of Solar Energy, "Development of a 40 kW Wind Turbine Generator," Federal Wind Energy Program; Request for Proposal No. PF64100F; March 28, 1977.

*Energy Research and Development Administration, "Federal Wind Energy Program, Summary Report," January 1, 1977, 56 pp., U.S. Government Printing Office (Stock No. 060-000-00048-4).

*Energy Research and Development Administration, "Federal Wind Energy Program, Summary Report," Division of Solar Energy, October 1975, 78pp. (ERDA-84).

*These reports are available from the National Technical Information Service, 5285 Port Royal Road, Springfield, VA 22161, (703) 557-4650 and/or The Superintendent of Documents, U.S. Government Printing Office, Washington, D.C. 20402, (202) 783-3238.

Fairbridge, R.W., ed., "The Encyclopedia of Atmospheric Science and Astrogeology," 1967.

Federal Energy Administration, "Project Independence Report on Solar Energy," November 1974 (Available from GPO).

*General Electric, Space Division, "Wind Energy Mission Analysis," February 1977, Contract No. E(11-1)-2578. Executive Summary: C00/2578-1/1, 2pp.; Final Report: C00/2578-1/2, 219 pp. Appendices A-J: C00/2578-1/3, 480 pp.

Golding, E.W., "The Generation of Elecricity by Wind Power," Philosophical Library, New York, 1956.

Institute for Energy Analysis, "Net Energy Analysis of Five Energy Systems," Oak Ridge Associated Universities, September 1977.

*JBF Scientific Corporation, "Summary of Current Cost Estimates of Large Wind Energy Systems," (Special Technical Report), February 1977, 62 pp., Contract No. E(49-18)-2364, (DSE/2521-1).

*JBF Scientific Corporation, "Third Wind Energy Workshop, Volumes 1 and 2," (September 19 to 21, 1977), Theodore R. Kornreich, ed., Contract No. E(49-18)-2521.

Kuchemann, D. and J. Weber, "Aerodynamics of Propulsion," McGraw-Hill, New York, 1953.

*Lockheed California Company, "Wind Energy Mission Analysis," October, 1976, Contract No. EY-76-C-03-1075, Executive Summary: SAN/1075-1/3, 30 pp.; Final Report: SAN/1075-1/1; Appendix: SAN/1075-1/2.

*MITRE Corporation, "Wind Energy Conversion Systems, Proceedings of the Second Workshop," (June 9-11, 1975), F.R. Eldridge, ed., June 1975, 536 pp., Contract No. NSF-AER-75-12937, (NSF-RA-N-75-050).

*MITRE Corporation, "Wind Machines," F.R. Eldridge, October 1975, 84 pp., (NSF-RA-N-75-051), U.S. Government Printing Office, (Stock No. 038-000-00272-4).

MITRE Corporation, "Wind Energy Resource Parameters," M.R. Gustavson, February 1977, M77-29, McLean, VA 22101.

*NASA-Lewis Research Center, "Wind Energy Conversion Systems, Workshop Proceedings," (Washington, D.C., June 11-13), J.M. Savino, ed., December 1973, 258 pp., Grant No. NSF AG465, (NSF-RA-N-73-006), (PB 231341).

*NASA-Lewis Research Center, "Wind Energy Utilization, A Bibliography," Technical Applications Center, University of New Mexico, for NASA LeRC, (TACW-75-700).

New Mexico State University, "A Wind Energy Summary for New Mexico," K.M. Barnett, Physical Science Laboratory, Box 3548, Las Cruces, NM 88003, January 1978.

*Oregon State University, "Applied Aerodynamics of Wind Power Machines," R.E. Wilson and P.B.S. Lissaman, May 1974, NTIS PB 238594.

*Ibid.

Portola Institute, "Energy Primer," R. Merrill, et al., eds., Menlo Park, California, 1974.

Putnam, P.C., "Power from the Wind," Van Nostrand Reinhold, 1948.
Torrey, Volta, "Wind-Catchers—American Windmills of Yesterday and Tomorrow," Stephen Greene Press, Brattleboro, VT 05301.

*United States Department of Energy, Division of Solar Technology, "Federal Wind Eergy Program: Program Summary," DOE/ET-0023/1, January 1978.

Windworks, "Wind Energy Bibliography," Box 329, Route 3, Mukwonago, WI 53149, 1974.

APPLICATIONS OF WIND ENERGY

Advisory Group for Aeronautical Research and Development, "The Influence of Aerodynamics in Wind Power Development," E.W. Golding, N63-80507, CFSTI (Available from NASA-Lewis).

*Michigan State University, Division of Engineering Research, "Application Study of Wind Power Technology to the City of Hart, Michigan," J. Asmussen, P.D. Fisher, G.L. Park, O. Krauss, December 1975, 103 pp., Contract No. E(11-1)-2603, (C00-2603-1).

*NASA-Lewis Research Center, "Benefit-Cost Methodology Study with Example Application of the Use of Wind Generators," R.P. Zimmer, C.G. Justus, S.L. Robinette, P.G. Sassone, W.A. Schaffer of Georgia Institute of Technology, July 1975, 411 pp., (NASA CR-134864).

*Technology Applications Center, "Wind Energy Utilization: A Bibliography," University of New Mexico, June 1975, PB 247-970.

U.S. House of Representatives, "Wind Energy," Hearings Before the Subcommittee on Energy of the Committee on Science and Astronautics, May 1974.

LEGAL/SOCIAL/ENVIRONMENTAL ISSUES

*Battelle Memorial Institute, Columbus Laboratories, "An Evaluation of the Potential Environmental Effects of Wind Energy Systems Development," (Final Report), S. Rogers et al., August 1976, Contract No. NSF-AER-75-07378 (ERDA/NSF/07378-75/1).

*George Washington University, "Legal-Institutional Implications of Wind Energy Conversion Systems," (Final Report), L.H. Mayo, et al., September 1977, 333 pp. Contract No. APR 75-19137, (NSF/RA-770204).

*Michigan, University of, "Radiation Laboratory TV and FM Interference by Windmills," (Final Report), T.B.A. Senior, et al., February 1977, 159 pp., Contract No. EY-76-S-02-2486, (C00/2846-76/1).

*Societal Analytics Institute, Inc., "Barriers of the Use of Wind Energy Machines: The Present Legal/Regulatory Regime and a Preliminary Assessment of Some Legal/Political/Societal Problems," R.F. and H.J. Taubenfeld, July 1976, 159 pp., Contract No. NSF-AER75-18362, (PB-263 567).

*Ibid.

WIND CHARACTERISTICS

*Alaska, University of, Geophysical Institute, "Study of Alaskan Wind Power and Its Possible Applications," (Final Report, May 1, 1974—January 30, 1976), T. Wentink, Jr., February 1976, 139 pp., Contract No. NSF-AER-74-00239 (NSF/RANN/SE/AER-74-00239) (PB 253 339).

*Alaska, University of, Geophysical Institute, "Wind Power Potential of Alaska, Part II, Wind Duration Curve Fits and Output Power Estimates for Typical Windmills," T. Wentink, Jr., August 1976, 86 pp., Contract No. E(45-1)-2229, (RLO/2229/T12-16/1).

*Battelle-Pacific Northwest Laboratories, "Annual Report of the Wind Characteristics Program Element for the Period April 1976–June 1977," J.V. Ramsdell, June 1977, Contract No. EY-76-C-06-1830, (BNWL-2220-WIND-10).

*Colorado State University, "Sites for Wind Power Installations: Wind Tunnel Simulation of the Influence of Two-Dimensional Ridges on Wind Speed and Turbulence," (Annual Report), R.N. Meroney, et al., July 1976, 80 pp., Con-1976, 129 pp., Contract No. NSF-AER-75-00547, (PB 260 679).

*Georgia Institute of Technology, "Reference Wind Speed Distributions and Height Profiles for Wind Turbine Design and Performance Evaluation Applications," C.G. Justus, W.K.Hargraves, A. Mikhail, August 1976, 96 pp., Contract No. E(40-1)-5108, (ORO/5108-76/4).

*Georgia Institute of Technology, "Wind Energy Statistics for Large Arrays of Wind Turbines (New England and Central U.S. Regions)," C.G. Justus, August 1976, 129 pp., Contract No. NSF-AER75-00547,, (PB 260 679).

Justus, C.G., "Wind Energy Statistics for Large Arrays of Wind Turbines (New England and the Central U.S. Regions)," Solar Energy, Vol. 20, No. 5, 1978, pp. 379-386.

*NOAA-National Climatic Center, "Initial Wind Energy Data Assessment Study," M.J. Changery, May 1975, 132 pp., Contract No. NSF-AG-517, (NSF-RA-N-75-020) (PB 244 132).

Reed, J.W., "Wind Power Climatology," Weatherwise, December 1974.

Sandia Laboratories, Contract No. S189-76-32:
  *"Wind Energy Potential in New Mexico," J.W. Reed, R.C. Maydew, B.F. Blackwell, July 1974, 40 pp., (SAND74-0071).

  *"Wind Power Climatology," J.W. Reed, December 1975 (SAND74-0435).

  *"Wind Power Climatology of the United States," J.W. Reed, May 1975, 163 pp., (SAND74-3078).

Western Scientific Services, Inc., "Candidate Wind Turbine Generator Sites," Monthly Reports prepared for NASA-Lewis Research Center, Clevelan ¹ OH, under Contract No. NAS3-20452.

*Ibid.

TECHNOLOGY DEVELOPMENT

*General Electric, Space Division, "Design Study of Wind Turbines 50 kw to 3000 kW for Electric Utility Applications," December 1976, Volume I (Summary Report): NASA CR-134934; Volume II: NASA CR-134935; Volume III: NASA CR-134936.

General Electric Company, "Applied Research on Energy Storage and Conversion for Photovoltaic and Wind Energy Systems," Final Report, HCP/T-22221-01/3, January 1978.

General Electric Company, "Large-Scale Thermal Energy Storage for Cogeneration and Solar Systems," G.E. Tempo, Center for Advanced Studies, Santa Barbara, CA, February 24, 1978.

*Hutter, U., "Operating Experience Obtained With a 100-kW Wind Power Plant," Kanner Associates, N73-29008/2, NTIS, 1964.

*Lockheed-California Company, "100 kW Metal Wind Turbine Blade Basic Data, Loads, and Stress Analysis," A.W. Cheritt and J.A. Gaidelis, June 1975, NASA Contract No. NAS3-19235, (NASA CR-134956).

*Lockheed-California Company, "100 kW Metal Wind Turbine Blade Dynamics Analysis, Weight/Balance and Structural Test Results," W.D. Anderson, June 1975, NASA Contact No. NAS3-19235, (NASA-CR-134957).

*Martin Marrietta Laboratories, "Segmented and Self-Adjusting Wind Turbine Rotors," (Final Report), P.F. Jordon, R.L. Goldman, April 1976, 113 pp., Contract No. EY-76-C-02-2613, (C00/2613-2).

*Massachusetts Institute of Technology, "Research on Wind Energy Conversion Systems," R.H. Miller, December 1976, Contract No. NSF-AER-75-00826.

MITRE Corp./Metrek Division, "Wind Energy Conversion Systems Using Compressed Air Storage," F.R. Eldridge, M76-39, July 1976.

NASA-Lewis Research Center, Interagency Agreement No. E(49-26)-1028:

*"Free Vibrations of the ERDA-NASA 100 kW Wind Turbine," C.C. Chamis, T.L. Sullivan, February 1976, (NASA TM X-71879).

*"Transient Analysis of Unbalanced Short Circuits of the ERDA-NASA 100 kW Wind Turbine Alternator," H.H. Hwang, Leonard J. Gilbert, July 1976, (NASA TM X-73459).

*"Early Operation Experience on the ERDA/NASA 100 kW Wind Turbine," J.C. Glasgow, B.S. Linscott, September 1976, (NASA TM X-71601).

*"Tower and Rotor Blade Vibration Test Results for a 100 Kilowatt Wind Turbine," B.S. Linscott, W.R. Shapton, D. Brown, October 1976, (NASA TM X-3426).

*"Wind Tunnel Measurements of the Tower Shadow on Models of the ERDA/NASA 100 kW Wind Turbine Tower," J.M. Savino and L.H. Wagner, November 1976, (NASA TM X-73548).

*Ibid.

*"Synchronization of the ERDA/NASA 100 kW Wind Turbine Generator with Large Utility Networks," H.H. Hwang and L.J. Gilbert, March 1977, 17 pp., (NASA TM X-73613).

*"Vibration Characteristics of a Large Wind Turbine Tower on Non-Rigid Foundations," S.T. Yee, T. Yung, P. Change, et al., May 1977, (ERDA/NASA 1004-77/1).

*"Dynamic Blade Loading in the ERDA/NASA 100 kW and 200 kW Wind Turbine," D.A. Spera, D.C. Janetzke, T.R. Richards, May 1977, (ERDA/NASA 1004-77/2).

*"Drive Train Normal Modes Analysis for the ERDA/NASA 100-Kilowatt Wind Turbine Generator," T.L. Sullivan, D.R. Miller, D.A. Spera, July 1977, (ERDA/NASA 1028-77/1).

*"Investigation of Excitation Control for Wind Turbine Generator Stability," V.D. Gebben, August 1977, (ERDA/NASA 1028-77/3).

*"Nastran Use for Cyclic Response and Fatigue Analysis of Wind Turbine Towers," C.C. Chamis, P. Manos, J.H. Sinclair, J.R. Winemiller, October 1977, 20 pp., (ERDA/NASA 1004-77/1004-77/3).

*Oklahoma State University, "Development of an Electrical Generator and Electrolysis Cell for a Wind Energy Conversion System," (Final Report, July 1, 1973-July 1, 1975), W. Hughes, H.J. Allison, R.G. Ramakumar, July 1975, 280 pp., Contract No. NSF-AER-75-00647, (NSF/RA/N-75-043) (PB 243 909).

*Oregon State University, "Applied Aerodynamics of Wind Power Machines," R.E. Wilson, P.B.S. Lissaman, July 1974, 116 pp., Contract No. NSF-AER-74-04014 A03 (PB 238 595).

*Oregon State University, "Aerodynamic Performance of Wind Turbines," R.E. Wilson, P.B.S. Lissaman, S.N. Walker, June 1976, 170 pp., Contract No. NSF-AR-74-04014 A03(PB 259 089).

*Paragon Pacific, Inc., "Coupled Dynamics Analysis of Wind Energy Systems," February 1977, NASA Contract No. NAS3-197707 (NASA CR-135152).

Swedish Board for Technical Development, "Wind Power in Sweden—A Preliminary Feasibility Study," O. Ljungstrom and B. Sodergard, May 1974.

*United Technologies Research Center, "Self-Regulating Composite Bearingless Wind Turbine," (Final Report, June 3, 1975–June 2, 1976), M.C. Cheney, P.A.M. Spierings, September 1976, 62 pp., Contract No. EY-76-C-02-2614, (C00/2614-76/1), Executive Summary, 13 pp., (C00/2614-76/2).

ADVANCED SYSTEMS

*AAI Corporation and Institute of Gas Technology, "Production of Methane Using Offshore Wind Energy," R.B. Young, A.F. Tiedeman, Jr., T.G. Marianawski, E.H. Camara, November 1975, Contract No. NSF-C993, Final Report: (PB 252 307), 131 pp., Executive Summary: (PB 252 308), 29 pp.

*Ibid.

*Dayton, University of, Research Institute, "Electrofluid Dynamic (EFD) Wind Driven Generator," J.E. Minardi, M.O. Lawson, G.Williams, October 1976, Contrct No. EY-76-S-02-4130.

*Grumman Aerospace Corporation, "Investigation of Diffuser-Augmented Wind Turbines," R.A. Oman, January 1977, Contract No. EY-76-C-02-2616, Executive Summary: (C00-2616-1), Technical Report: (C00/2616-2).

Yen, J.T., "Tornado-Type Wind Energy System," Tenth Intersociety Energy Conversion Engineering Conference, Catalog No. 75 CHO 983-7 TAB, August 1975.

*McDonnell Aircraft Co., "Feasibility Investigation of the Gyromill for Generation of Electrical Power," R.V. Brulle, November 1975, 155 pp., Contract No. E(11-1)-2617 (C00-2617-75/1).

*Princeton University, "Optimization and Characteristics of a Sailwing Windmill Rotor," M.D. Maughmer, March 1976, Contract No. GI-41891. (NSF/RANN/GI-41891/FR/75/4).

Sandia Laboratories, Contract No. AT(29-1)-789:

  *"Vertical-Axis Wind Turbine Technology Workshop," (held at Sandia Laboratories, Albuquerque, New Mexico, May 18-20, 1976), L. Wetherholt, ed., July, 1976, 439 pp., (SAND76-5586).

  *"The Vertical-Axis Wind Turbine - How it Works," B.F. Blackwell, April, 1974, 8 pp., (SLA-74-0160).

  *"Blade Shape for a Troposkien Type of Vertical-Axis Wind Turbine," B.F. Blackwell, G.E. Reis, April 1974, 24 pp, (SLA-74-0154).

  *"An Electrical System for Extracting Maximum Power from the Wind," A.F. Veneruso, December 1974, 29 pp. (SAND74-0105).

  *"Some Geometrical Aspects of Troposkiens as Applied to Vertical-Axis Wind Turbines," B.F. Blackwell, G.E. Reis, March 1975, (SAND74-0177).

  *"Practical Approximations to a Troposkien by Straight-Line and Circular-Arc Segments," G.E. Reis, B.F. Blackwell, March 1975, 34 pp. (SAND74-0100).

  *"Wind Energy—A Revitalized Pursuit," B.F. Blackwell, L.V. Feltz, March 1975, 16 pp. (SAND75-0166).

  *"An Investigation of Rotation-Induced Stresses of Straight and of Curved Vertical-Axis Wind Turbine Blades," L.V. Feltz, B.F. Blackwell, March 1975, 20 pp. (SAND74-0379).

  *"Application of the Darrieus Vertical-Axis Turbine to Synchronous Electrical Power Generation," J.F. Banas, E.G. Kadlec, W.N. Sullivan, March 1975, 14 pp. (SAND75-0165).

  *"Nonlinear Stress Analysis of Vertical-Axis Wind Turbine Blades," W.I. Weingarten, R.E. Nickell, April 1975, 20 pp. (SAND74-0378).

  *"Methods of Performance Evaluation of Synchronous Power Systems Utiliz-

*Ibid.

ing The Darrieus Vertical-Axis Wind Turbine," J.F. Banas, E.G. Kadlec, W.N. Sullivan, April 1975, 22 pp. (SAND75-0204).

*"The Darrieus Turbine: A Performance Prediction Model Using Multiple Streamtubes," J.H. Strickland, October 1975 (SAND75-0431).

*"Engineering of Wind Energy Systems," J.F. Banas, W.N. Sullivan, January 1976 (SAND75-0530).

*"Wind Tunnel Performance Data for the Darrieus Wind Turbine with NACS-0012 Blades," B.F. Blackwell, L.V. Feltz, R.E. Sheldahl, 1976 (SAND76-0130).

*"Wind Tunnel Performance Data for Two- and Three-Cup Savonius Rotors," B.F. Blackwell, L.V. Feltz, R.E. Sheldahl, July 1977, 108 pp. (SAND76-0131).

*"Engineering Development Status of the Darrieus Wind Turbine," B.F. Blackwell, W.N. Sullivan, R.C. Reuter, J.F. Banas, March 1977, 68 pp. (SAND76-0650).

*"Darrieus Vertical-Axis Wind Turbine Program at Sandia Laboratories," E.G. Kadlec, August 1976, 11 pp. (SAND76-5712).

*"Status of the ERDA/Sandia 17-Meter Darrieus Turbine Design," B.F. Blackwell, September 1976, 16 pp. (SAND76-5683).

*West Virginia University, "Innovative Wind Machines," (Executive Summary and Final Report), R.E. Walters, et al., June 1976, Contract No. EY-76-C-05-5135 (ERDA/NSF/00367-76/2).

SMALL-SCALE SYSTEMS

*Colorado State University, "Wind-Powered Aeration for Remote Locations," (Final Report, March 15, 1975–August 31, 1976), P.M. Schierholz, October 1976, 130 pp., Contract No. NSF-G-AER-75-00833, (ERDA/NSF/00833-75-1).

Enertech Corporation, "Planning a Wind-Powered Generating System," Box 420, Norwich, VT 05055, February 1977.

*Institute of Gas Technology, "Wind-Powered Hydrogen-Electric Systems for Farm and Rural Use," J.B. Pangborn, April 1976, 158 pp., Contract No. NSF-AER-75-00772 (PB 259 318).

*Massachusetts University, Amherst, "Investigation of the Feasibility of Using Windpower for Space Heating in Colder Climates," (Third quarterly progress report covering the final design and mfg. phase of the project, September–December, 1975), W.E. Heronemus, December, 1975, 165 pp., Contract No. NSF-AER-75-00603 (ERDA/NSF/00603-75/T1).

MITRE Corporation/Metrek Division, "A Survey of Small-Scale, Water-Pumping Wind Machines," F.R. Eldridge, WP-12546, August 1977.

---

*Ibid.

*NASA-Lewis Research Center, "Installation and Initial Operation of a 4100 Watt Wind Turbine," H.B. Tryon, T. Richards, December 1975 (NASA TM X-71831).

National Academy of Science, "Energy for Rural Development," Washington, D.C., 1976.

Nielsen Engineering and Research, Inc., "Wind Power for Farms, Homes and Small Industry," Jack Park and Dick Schwind, 510 Clyde Avenue, Mountain View, CA 94043, January 1977.

Park, J., "Simplified Wind Power Systems for Experimenters," Helion Inc., Box 4301, Sylmar, CA 91342, 1975.

Solar Wind Co., "Electric Power from the Wind," H. Clews, P.O. Box 420, Norwich, VT 05055, 1974.

Total Environmental Action, "An Introduction to the Use of Wind," Douglas Coonley, Church Hill, Harrisville, NH 03450, May 1974.

Total Environmental Action, "Design with Wind," Douglas Coonley, Church Hill, Harrisville, NH 03450, May 1974.

100 KILOWATT-SCALE SYSTEMS

NASA-Lewis Research Center:

    *"Preliminary Design of a 100 kW Turbine Generator," R.L. Puthoff, P.J. Sirocky, 1974, 22 pp. (NASA TM X-71585; E-8037) (N-74-31527).

    *"Structural Analysis of Wind Turbine Rotors for NSE-NASA MOD-0 Windpower System," D.A. Spera, March 1975, 39 pp. (NASA TM X-3198; E-8133) (N-75-17712).

*"Plans and Status of the NASA-Lewis Research Center Wind Energy Project," R. Thomas, R. Puthoff, J. Savino, W. Johnson, 1975, 31 pp. (NASA TM X-71701; E8309) (N-75-21795).

*"A 100 kw Experimental Wind Turbine: Simulation of Starting Overspeed and Startdown Characteristics," L. Gilbert, February 1976, (NASA TM X-71864).

*"Large Experimental Wind Turbines—Where We Are Now," R.L. Thomas, March 1976 (NASA TM X-71890).

*Fabrication and Assembly of the ERDA/NASA 100 kW Experimental Turbine," R.L. Puthoff, April 1976 (NASA TM X-3390).

*"Design Study of Wind Turbines 50 kW to 3000 kW for Electric Utility Applications," (See Technology Development).

MEGAWATT-SCALE SYSTEMS

*General Electric, Space Division, "Design Study of Wind Turbines 50 kW to 3000 kW for Electric Utility Applications," (See Technology Development).

*NASA-Lewis Research Center, "Large Experimental Wind Turbines," (See 100 kW-Scale Systems).

---

*Ibid.

Tvind Schools, "The Wind Machine at Tvind," Ingenioren, No. 26, June 25, 1976, Reprint, Available from MITRE/Metrek, McLean, VA 22101.

COMMERCIALIZATION

American Wind Energy Association, "The Federal Wind Program: A Proposal for FY 1979 Budget," prepared for The Department of Energy, 18 February 1978.

Council on Environmental Quality, "Solar Energy—Progress and Promise," Executive Office of the President, 722 Jackson Place, N.W., Washington, D.C., April 1978.

*Interior, U.S. Department of the, "Cost-Effective Electric Power Generation from the Wind," C.J. Todd, et al, Division of Atmospheric Water Resources Management, Engineering and Research Center, Bureau of Reclamation, August 1977, PB 273 582.

*Lawrence Berkeley Labs, Lawrence Livermore Labs and University of California (Berkeley), "Distributed Technology in California's Energy Future," Report to the Office of Environmental Policy Analysis, ERDA, LBL 6831, September 1977.

Lockheed California Company, "Effects of Initial Production Quantity and Incentives on the Cost of Wind Energy," U. Coty and L. Vaughn, Burbank, CA January 1977.

MITRE Corporation/Metrek Division, "Preliminary Federal Commercialization Plan for Wind Energy Conversion Systems," F.R. Eldridge, et al., MTR-7365, January 1977.

MITRE Corporation, "Commercialization of Small-Scale Wind Machimes," Presentation to the U.S. House of Representatives Ways and Means Committee, F.R. Eldridge, M77-16, June 1977.

MITRE Corporation/Metrek Division, "A Typical Business and Marketing Plan for an Energy-Related Small Business," F.R. Eldridge, MTR-7651, October 1977.

MITRE Corporation/Metrek Division, "Solar Energy—A Comparative Analysis to the year 2020," MTR-7579, McLean VA, March 1978.

SRI-International, "Solar Energy Research and Development—Program Balance," HCP/M2693-01, Prepared for U.S. Department of Energy, Solar Working Group, Under Contract No. EA-77-C-01-2693, February 1978.

Office of Technology Assessment, "Application of Solar Technology to Today's Energy Needs," Washington, D.C., March 1978.

JOURNALS

Mother Earth News, The, P.O. Box 70, Hendersonville, NC 28739.

"Newsletter," published by Wind Energy Society of America, 1700 East Walnut St., Pasadena, CA 91106.

*Ibid.

Popular Science, published by Times Mirror Magazines, Inc., 380 Madison Ave., New York NY 10017.

Solar Age, published by SolarVision Inc., Church Hill, Harrisville, NH 03450.

Solar Energy, published by Pergamon Press for International Solar Energy Society, P.O. Box 52, Parkville, Victoria, Australia 3052.

Wind Energy Report, PO Box 14, Rockville Center, New York 11571.

Wind Engineering, published by Multi-Science Publishing Co. Ltd., The Old Mill, Dorset Place, London E15 1DJ, England.

"Windletter," published by American Wind Energy Assn., Suite 1111, 1717 K St., N.W., Washington, D.C.

Wind Power Digest, published by Michael Evans, 54468 CR31, Bristol, IND 46507.

Wind Technology Journal, published by American Wind Energy Assn., P.O. Box 7, Marstons Mills, MA 02648.

"Windustries," published by Great Plains Windustries, Box 126, Lawrence, KA 66044.

*Ibid.

INDEX